GLOBAL TAIWANESE

Global Taiwanese

Asian Skilled Labour Migrants in a Changing World

FIONA MOORE

UNIVERSITY OF TORONTO PRESS
Toronto Buffalo London

© University of Toronto Press 2021
Toronto Buffalo London
utorontopress.com

ISBN 978-1-4875-0001-6 (cloth)
ISBN 978-1-4875-1000-8 (EPUB)
ISBN 978-1-4875-0999-6 (PDF)

Library and Archives Canada Cataloguing in Publication

Title: Global Taiwanese : Asian skilled labour migrants in a changing world / Fiona Moore.
Names: Moore, Fiona, author.
Description: Includes bibliographical references and index.
Identifiers: Canadiana (print) 20210090545 | Canadiana (ebook) 2021009074X | ISBN 9781487500016 (cloth) | ISBN 9781487510008 (EPUB) | ISBN 9781487509996 (PDF)
Subjects: LCSH: Taiwanese – England – London. | LCSH: Taiwanese – Ontario – Toronto. | LCSH: Taiwanese – Taiwan – Taipei. | LCSH: Taiwanese – Ethnic identity. | LCSH: Social networks.
Classification: LCC DS799.42 .M66 2021 | DDC 304.8089/9925 – dc23

University of Toronto Press acknowledges the financial assistance to its publishing program of the Canada Council for the Arts and the Ontario Arts Council, an agency of the Government of Ontario.

Canada Council for the Arts
Conseil des Arts du Canada

Funded by the Government of Canada
Financé par le gouvernement du Canada
Canada

For my parents

Contents

Acknowledgments

Given that, to overuse a cliché, behind every ostensibly single-authored book is a whole community of supporters, the author would like to acknowledge the help of, in no particular order, Professor Maria Jaschok of the Oxford China Centre and the Women and Gender in Chinese Studies Network for support and general guidance, Doctor Yih-teen Lee for reading and critiquing the data chapters, Doctor Chin-Ju Tsai of Royal Holloway University of London for the initial introduction to the London Taiwanese community; Tracy Hsu, Leslie Lin and her family, and Bronwen Moore and her family, for introducing me to the Toronto Taiwanese community; Robert and Jane Moore for support, accommodation and occasional transport assistance during the Toronto phase of the study; Vicky Li for linguistic support; Doctor Dorothy Yen, Polly Lee, and Doctor Jean Wang for providing me with new insights into the Taiwanese community in London; Professor Rueylin Hsiao for helping me to get to Taiwan; the administrative and assistant team at National Cheng Chi University's College of Commerce for support, guidance and translation while I was there; the staff and members of the various Taiwanese community, business, and consular organizations who participated in this study; and Jennifer DiDomenico and the anonymous peer reviewers at the University of Toronto Press for editorial support and assistance.

Initial financial support for the London side of the project was provided by the Nuffield Foundation. The Toronto portion of the study was supported by a small grant from Royal Holloway University of London's Critical and Historical Perspectives Research Group, and National Cheng Chi University provided me with a visiting fellowship and financial support to conduct research in Taiwan itself. All this assistance was gratefully received and the study could not have been conducted without it.

Glossary of Terms and Acronyms

Benshengren "This province people," the descendants of the original Chinese colonists of Taiwan

ETCC European Taiwanese Chamber of Commerce, regional level of representation of Taiwanese chambers of commerce

DPP Democratic Progressive Party, centre-leftist party in Taiwan

Guanxi A form of networking through mutual and reciprocal exchanges of favours

NCCU National Cheng Chi University

NTU National Taiwan University

KMT/Kuomintang "Republican Party" or "Chinese Nationalist Party," a political party founded in 1919, which took power in Taiwan following the 1949 migration from the mainland and remained in power until 2000; currently a centre-right party in Taiwan

Taipei Representative Offices Consular organizations representing Taiwan in countries that do not officially recognize Taiwan as an independent state

TAITRA Taiwanese Trade Association, an organization representing Taiwanese businesses in the UK

TMAT Taiwanese Merchants' Association of Toronto

Waishengren "Outside province people," Chinese people who migrated to Taiwan from the mainland in 1949, and their descendants

WTCC World Taiwanese Chamber of Commerce, international body of Taiwanese chambers of commerce

GLOBAL TAIWANESE

Chapter One

Why Taiwan? Taiwanese Identity and the Chinese Diaspora

While conducting the study on which this book is based, the most common reaction I heard from people – including from Taiwanese people – was "That's interesting. So, why Taiwan?" In this introductory chapter, I explain why Taiwan, a dynamic economic and political power emerging from the shadow of mainland China and exploring its complicated relationship with its larger neighbour, is worth considering, both in and of itself and for the wider lessons it can teach us about social and economic organization in a technically oriented, networked, communication-focused world in which a growing need for skilled labour migrants exists in tension with increasingly nationalist migration policies on the part of local governments.

This book will explore both these issues by describing the lives and opinions of two groups of Taiwanese skilled labour migrants and the infrastructures that support them, examining the ways they actively use their identities as Taiwanese, as Chinese, and as transnational professionals as tools for building connections and pursuing business activities in Taiwan, Europe, and North America. To that end, I will begin by documenting and exploring the perspective of a unique group emerging as a new technological and economic power in Asia, one that has a particularly complex and contested identity, developed in the shadow of a powerful and threatening neighbour, and that presents a new and surprising approach to transnational and Chinese identities. I will then address the limits of analysis with respect to ethic and nationalist identity. After that, I will develop the concept of the "network society" as a starting point for considering the lived implications of seeing a diaspora, not in terms of home countries and countries of migration, but as nodes in a network, one in which different concepts of being local and global are experienced and reinterpreted by skilled labour migrants. The idea of the "network society" has been around for twenty years (see, for instance,

Castells, 1996, 1997/2004, 1998/2010), but most post-millenial studies examining life in this situation focus on migrants at one node or another on their journey, rather than looking at multiple nodes in the same network at more or less the same time; I would also argue that the concept can be expanded to consider whether and how transnational networks have evolved in the generation since they were developed. Finally, I will explore the social infrastructure through which this activity persists and develops, such as business support groups and online communities, and how they enable skilled labour migrants to operate across borders even in contexts in which legal and social barriers attempt to block them from doing so; in the process, I will be problematizing the concepts of ethnic, family, and professional networks.

This volume is, therefore, not only about international skilled labour migration and the Chinese/Taiwanese diaspora, and the rise of Taiwan on the world stage, but also about the changing nature of transnational living and networking across borders, how earlier concepts of the network society and transnational ethnic identities have developed in light of later geopolitical trends, and what we can learn from the stories of the people who live in a globe-spanning network society. First, however, we must explore what makes Taiwan a potentially significant starting point for this project.

A Free China? Taiwanese History and Background

Chapters 4, 5, and 6 offer more detailed accounts of the communities in London, Toronto, and Taipei respectively. A full account of Taiwan's fascinating history is beyond the scope of this particular book (for more detailed histories, I would recommend, among many others, Liu & Hung [2002], Davidson [1903/2007], Andrade [2007], Harrison [2007], Wu [2004], Wong [2004], and Dawley [2009]). Yet it is impossible to understand the complexities of Taiwanese identity without considering how it has developed over the past five hundred years. For space reasons, I will focus on those aspects of history that my informants mentioned as being particularly relevant to their experiences and on those which, based on my own participant-observation, I believe are the most helpful in understanding the Taiwanese experience today.

A Brief and Idiosyncratic History of Taiwan

Taiwan is a large island off the southwest coast of mainland China. Its original settlers were aboriginal groups, who had some contact with the Chinese mainland from an early period. The Chinese began visiting regularly and established settlements on the island no later than the Yuan dynasty (1271–1368) (Andrade, 2007, p. 268). It was the Portuguese, in 1544, who first brought the island to European attention (Andrade, 2007, ch. 1). However, serious efforts to colonize Taiwan did not begin until the seventeenth century. These were initiated by the Dutch – with a brief period of Spanish occupation in the island's north – followed by the Chinese (Andrade, 2007). Taiwan remained a colony of China until 1895, becoming a full province in 1886 (Liu & Hung, 2002, p. 569), although Hall (2003, p. 154) notes that the mainland paid little attention to Taiwan for most of this period. Ethnically, the population was (as it still is) dominated by Han Chinese; other significant ethnic groups included the Hakka, the Fujian, and a number of aboriginal groups.

As part of the resolution of the Sino-Japanese War, Taiwan was ceded to Japan until 1945, a period that historians generally describe as one of social and infrastructural development for Taiwan: "Under Japan's firm control of the island ... intra-Han ethnic tensions eased ... Japan improved public safety, health, education, and the industrial and communication infrastructures. These changes created a Taiwanese middle class, many of whom remember the Japanese colonial period with nostalgia. Even working-class and rural Taiwanese remember the Japanese as strict but fair" (Brown, 2001, p. 154). Nostalgia for this period can still be observed in modern Taipei; although most official museums and at least one Confucian temple presented the Japanese rule as oppressive, all of the people I spoke with in Taiwan were generally positive toward it, pointing out Japanese traits in the city's architecture and taking an interest in Japanese high and popular culture. When I asked one of my Taiwanese research assistants why the Japanese period was so well regarded, she thought for a moment and then said it was because they established schools and hospitals. The Japanese occupation still plays a role in contemporary Taiwanese identity, negative for some and positive for others, but acknowledged by all.

After the Second World War, the defeated Japanese ceded Taiwan back to what was then Republican China (Liu & Hung, 2002, p. 570). This was – unsurprisingly, given the above-mentioned prosperity under Japanese rule – not a popular move with the local people, who felt that the mainland Chinese regarded them as second-class citizens (ibid.). Tensions were exacerbated when the mainland Chinese, according to Bedford and Hwang (2006, p. 6), largely dismantled the local infrastructure and shipped it to the mainland. This fuelled nostalgia for Japanese rule among the Taiwanese. The situation culminated in an insurgency in 1947 (Liu & Hung, 2002, p. 570).

In the aftermath of the civil war on the mainland, top members of the Kuomintang (KMT, loosely translatable as the "Republican Party") fled to Taiwan, along with a wave of their supporters, as well as others, such as businesspeople, whose objections to Communist rule were less political. In Taiwan, they and their descendants are still referred to as *waishengren*, a term meaning "outside province people" or, more loosely, "people who come from outside," in contrast to the *benshengren*, "this province people" or, more loosely, "people who come from here" – that is, the descendants of the original Chinese colonists (Hall, 2003, p. 145). Liu and Hung (2002, p. 570) describe their migration: "Nearly two million refugees, including government officials and military personnel, came to Taiwan ... In order to bolster its authority and legitimacy, the KMT constructed a political system designed to symbolize Taiwan's status as simply a province of the ROC, and to foster Chinese consciousness among the local population."

The KMT established the One China policy, which Liu and Hung summarize as follows: "(1) Taiwanese are Chinese. (2) Taiwan is a part of China. (3) The government of the ROC on Taiwan has to maintain a political system which can represent the whole of China" (570). The country was placed under martial law, which was not lifted until 1987. Throughout those years, Taiwan was a dictatorship, first under Chiang Kai-Shek (Jiang Jieshi) and later under his son, Chiang Ching-Kuo (571). The nostalgia for Japanese rule my informants expressed can be viewed as a subtle expression of anti-KMT sentiment and, at times, *benshengren* identity.

During the years it exercised martial law, the KMT strongly supported a discourse of identity that presented Taiwan as the "real China," a kind of pure Chineseness that was untainted by communism (Hall, 2003, p. 135; Murray & Hong, 1991) and was the historical continuation of Imperial and post-Imperial China (Chen, 2012, p. 853). Murray and Hong (1991) write that this had a distinct effect on anthropological studies of the region: anthropologists keen to further romantic notions of a "pure" or "traditional" China were financially supported by the state for the sake of promoting the KMT's own image of Taiwan as retaining the genuine culture of China (see also Liu & Hung, 2002; Harrison, 2007). Murray and Hong (1991) write that the KMT dictatorship tended to restrict access to researchers who did not favour their point of view that Taiwan was not an independent nation but was instead Chinese (pp. 275–276). They add that this was abetted by a tendency among Western Sinologists to romanticize Taiwan as a kind of unspoiled repository of pre-Communist Chinese culture (p. 283), as well as by a reluctance on the part of US governments in the 1950s and 1960s to acknowledge a legitimate Chinese identity on the Communist mainland. The United States supported Taiwan's claim to true Chinese identity as a means to further the

Cold War against the People's Republic of China, at least until the Nixon administration established a detente with the PRC in the 1970s (p. 280), even though Taiwan has been under official Chinese control for only two rather brief periods in its history. US investment during this period, meanwhile, led to a situation in which Taiwan gained a reputation as a producer of cheap household goods, with "Made in Taiwan" entering the North American vernacular as shorthand for poorly made, low-quality goods (much, it is worth noting, as "Made In Japan" had been during the 1950s).

This shifting political situation whereby Taiwan was caught between Chinese interests on the one hand, and US interests on the other, led to what Murray and Hong (1991) called the "invisibility" of Taiwan in the literature, with much research being subsumed under the label of "China." For instance, Wolf's 1972 ethnographic work *Women and the Family in Rural Taiwan* was preceded by the same anthropologist's 1968 ethnography *The House of Lim: A Study of a Chinese Farm Family*; the site referred to as "Taiwan" in the later book had been described as "Chinese" in the earlier one (Harrison, 2007, p. 4). Castells notes that as late as the 1990s, much of the literature on "Chinese" business was actually based on studies of Taiwanese businesses (1996, p. 193). One reason for the relative scarcity of literature about Taiwan is, therefore, that until recently it was seldom acknowledged to exist as something other than "China," and its identity has been often subsumed into that of Chinese culture more generally.

The recognition of mainland China by the United States in the 1970s threw the KMT into something of an identity crisis, which it weathered by resorting to symbols such as a fondness for baseball to reaffirm the identity of Taiwanese as "Free Chinese" (Yu & Bairner, 2008). This focus on a "Free Chinese" identity was also embraced by individuals during the dictatorship: Hall says that many of his informants "did not begin to think of themselves as Taiwanese until political changes in Taiwan made it possible for them to more openly discuss issues of ethnic identity" (Hall, 2003, p. 160). However, some challenges to the KMT discourse were beginning to be mounted: for instance, Appleton's (1970) study found sharp differences between *waishengren* and *benshengren*, including differences related to class, location, and political attitudes. These worked against the idea that Taiwan had a single "Chinese" or "Free Chinese" identity, suggesting that this discourse was not always unchallenged or free of subversion.

The end of martial law in the 1980s saw the re-emergence of a distinctive Taiwanese identity and culture (Harrison, 2007, ch. 7). An alternative political party, the Democratic Progressive Party (DPP), was founded, which focused on Taiwanese independence from the mainland as a counter-discourse to the KMT's support for reunification. The DPP's stance implicitly legitimized Taiwan's existence as a distinctive culture in its own right rather than a museum-like repository of pre-Communist Chineseness (Bedford & Hwang, 2006; Liu & Hung, 2002). This rise of Taiwanese self-awareness continued through the 1990s, with Liu and Hung writing that

> according to a recent investigation, most people in Taiwan have a sense of dual identity or mixed identity, describing themselves as "both Chinese and Taiwanese." This dual identity is expressed in two different ways. Some people perceive themselves as Chinese in the cultural domain and as Taiwanese in the political domain. Some accept that Taiwan is geographically part of China, but nonetheless possess a strong sense of a distinctive "Taiwaneseness." (p. 572)

Taiwan's complicated history of colonization, and of conflict over which discourses of identity should dominate, has resulted in a nation with a complex local identity, which, besides having its own distinctive traditions, has contributed to discourses on Chinese identity in the wider world. As we will now see, this has placed Taiwan in a position where two competing identities coexist, both at home and in the diaspora.

Taiwan at the Time of the Study

The population of Taiwan, at 2016 estimates, stands at about 23.4 million (source: www.cia.gov/library/publications/the-world-factbook/geos/tw.html). The size of the diaspora is difficult to estimate, for most available statistics conflate it with the Chinese diaspora. However, according to my informants at the European Chamber of Commerce Taiwan, were 113 Taiwanese economic and cultural offices overseas as of mid-2017 (source: www.roc-taiwan.org/portalOfDiplomaticMission_en.html) and all major (and most minor) cities in both the developed and the developing world had a Taiwanese Chamber of Commerce. Academic accounts tend to conflate the Taiwanese and Chinese diasporas; that said, articles such as those by Chen (2004), Yu and Bairner (2008), Montgomery (2008), and Lien (2011) attest to the reach and influence of Taiwanese migrants specifically.

In the twenty-first century, Taiwan has largely shaken off its past image as a producer of cheap consumer goods and developed a reputation for quality products of various kinds, particularly electronics sector but in numerous others as well. (I remember, in the 1990s, seeing a percussion set whose woodblock bore a prominent label reading "Made With

Pride in Taiwan," arguably marking a transition period between the two eras.) Along with South Korea, Hong Kong, and Singapore, Taiwan is recognized as one of East Asia's "tiger economies," which have surged to prominence since the end of the Cold War, and a number of Taiwanese-born electronics firms, such as Acer and HTC, have gained international acclaim. The United States does not officially recognize Taiwan's independence, but the 1979 Taiwan Relations Act guarantees some respect for its economic and political sovereignty, and many US politicians are sympathetic to the idea of an independent Taiwan (*The Economist*, 2016b). It seems that in some ways, Taiwan is forging its own identity in the globalized world.

Other developments, however, indicate that Taiwan's identity continues to be divided. The rise of China as an economic power since the 1970s, and its dominance of the Asian market since the Japanese crisis of the late 1990s, as well as increasing US economic investment on the mainland, have together meant that the relationship between China and Taiwan continues to be fraught, with mainland China continuing to pursue its claim to Taiwan with varying degrees of aggressiveness, and with the desirability of China as an economic partner paradoxically strengthening the KMT's contention that Taiwan is a Chinese country, since a claim to Chineseness legitimates economic closeness to the mainland (an irony that will be unpacked in later chapters). Taiwan is thus presently caught between the allure of China on the one hand and its own status as a burgeoning East Asian economy on the other.

Although *waishengren* continue to dominate mainstream political discourse, there have been recent challenges from *benshengren* as the value of an independent Taiwanese identity gains currency (see, for instance, Bedford & Hwang, 2006; Hall, 2013; Muyard, 2012). Significantly, however, both groups are reported as self-identifying as "Taiwanese" rather than one group identifying as "Taiwanese" and the other as "Chinese," even if both groups may define Taiwanese identity in subtly different ways (see, for instance, Chow, 2012; Danielsen, 2012; Schubert, 2004). At the start of my fieldwork, the KMT dominated the government and was pursuing closer ties with China, in line with its historical position that Taiwan is the repository of true Chinese identity. Over the course of the study, the KMT waned, and a DPP-dominated government came to power in 2016 (*The Economist*, 2016a). This political shift was reflected in how people discussed Taiwanese identity in interviews and informal conversations, and also in more subtle ways, such as in terms of the priorities and activities of trade-promoting organizations, and the performances and participants at community events I attended.

During my research, I could discern differences between *waishengren* and *benshengren* among the participants, although these were often not immediately obvious. Various markers of difference included political support for the KMT or the DPP and/or whether they favoured reunification with the mainland. This tallied with Bedford and Hwang's account of the differences between the groups (2006, ch. 1), in which they characterized the *waishengren* as supporters of the KMT and of reunification with China, and the *benshengren* as more focused on crafting an independent identity for Taiwan. *Benshengren* also tended to look to the Japanese occupation as a source of identity, using a Japanese education or older relatives who speak Japanese as an identity marker. The division between the two groups thus continues, in subtler form, to the present day.

However, this seemingly simple differentiation could be complicated in practice. For instance, over the course of the study I encountered people of *waishengren* origin who were not interested in reunification, sympathizing instead with the *benshengren*. Even *waishengren* identity could be less than straightforward: my study inclued some people who were technically classed as *waishengren* by virtue of their families having migrated from the mainland in 1949, but who were different from the KMT supporters in that they had fled the Communist regime for other social, political, or economic reasons – for instance,one woman's father had migrated to Taiwan to prevent the Communist regime from seizing his business. I also encountered *benshengren* who were not particularly concerned whether Taiwan remained independent of mainland China or not. I would describe Taiwanese identities less as "divided" than as focused on debates and discourses with their roots in recent Taiwanese and Chinese history.

In addition, I often found my interviewees somewhat ambivalent about whether to identify as "Chinese," an example being this exchange at gathering sponsored by the European Taiwanese Chamber of Commerce: "I take the occasion to mention the Young Chinese Professionals in Scotland online network, and he looks a bit glum and says that the big question in Taiwan is whether Taiwan is part of China or something else. I say I thought 'the mainland left Taiwan,' using the phrase [KMT members] use, and he says he could talk about it a long time, but will leave it till later" (fieldnote, May 2010). I never learned whether the speaker was *waishengren* or *benshengren*; however, it is worth noting that the organization at the time had a strong KMT presence and that I was there at the invitation of a KMT member.

Conversely, it is possible to simultaneously maintain a Taiwanese *and* a Chinese identity, whether in the sense that particular groups within China have their own identities while also being Chinese (e.g. the Han, the Fujien, the citizens of Hong Kong), or in the sense that ethnically Chinese citizens of non-Chinese countries such as Singapore, Malaysia, the Caribbean, and Canada may consider themselves Chinese. Singapore, a Southeast Asian country that is ethnically dominated by the Chinese but is emphatically *not* part of China, might serve as a model for what an independent Taiwanese identity might be like. Chen (2012), exploring Taiwanese national identity, argues for this sort of simultaneous, complex identity, referring to it as "multifaceted" and "two-dimensional." Present-day Taiwanese may identify as Chinese in some ways and in some circumstances, but their doing so defies simple explanation.

Taiwanese identity remains complex and contested, hinging to some extent on whether the island should be considered a distinct country with a mixed albeit Chinese-dominated population, or a province (in one form or another) of China (in one form or another). Within this, of course, the situation is much more complicated: Harrison (2007) describes Taiwanese identity as "elusive" (p. 6), but also writes that "simply naming the island Taiwan ... constitutes a Taiwanese identity. Each time the island is named Taiwan and not Formosa or China, Taiwan as a legitimate object of meaning is being differentiated from other meanings that can encompass the island under other names" (p. 5). Taiwanese identity, how to define it, and its relationship with Chinese identity, is thus ambivalent and shot through with political and social questions beyond the immediate issues of personal identity (see Wang, 2004). Next I explore the implications of this for studying the Taiwanese diaspora.

The Taiwanese Diaspora

The Taiwanese diaspora has been little studied as diasporas go, partly because it is relatively small and partly because, as I noted earlier, researchers have often historically considered it as part of the Chinese diaspora (or, indeed, as actually *constituting* the Chinese diaspora) rather than as an entity in and of itself (Yang & Chang, 2010, Lo, 2002). The identification of a specifically Taiwanese diaspora would have been discouraged during the dictatorship years and perhaps was problematic for some time afterward (see Hall, 2003). To add to the complications, Taiwanese of Chinese origin are often considered part of the Han, Fujian, Hakka, or other Chinese diasporas instead, making the Taiwanese diaspora doubly diasporic (Lo, 2002, Wang, 2007). Other studies consider Taiwan as a node within the wider Chinese diaspora rather than as having its own diaspora (e.g., Yu & Bairner, 2008). Taiwan's historical "invisibility" (Murray & Hong, 1991), and the complex relationship between Taiwanese and Chinese identity, have together meant that its diaspora has, in the past, not been studied as its own phenomenon.

More recent studies have gone some way toward filling this gap in the literature. Sun (2014), for instance, looks at the return migration of aging Taiwanese from the United States, albeit in the context of the problematic health care situation in that country and the actions migrants to the United States take to obtain affordable care in their old age, and Montgomery (2008) considers the transnational activities of Taiwanese migrants in the hi-tech sectors as they construct useful social networks. Capellini and Yen (2015) look at how Taiwanese families in the United Kingdom use food as an identity marker. A number of studies of Chinese media, such as television and *wuxia* films, consider Taiwan and its diaspora as a nexus of media production *and* consumption (e.g., Funnell, 2014). Lan and Wu (2016) examine the distinctive identity construction of Taiwanese students in mainland China, and Tseng (2017) considers the personal, social, and political forces affecting the location decisions of mixed Chinese–Taiwanese couples who married in a third country. That the Taiwanese have a distinctive labour diaspora – indeed, a highly skilled one – is an emerging concept in the literature, even if such studies are still relatively few in number.

Unsurprisingly, research on the Taiwanese diaspora often engages with the complexity of Taiwanese identity and its connections to wider Chinese identities. Some consider Taiwanese identity as the source of an alternative form of Chineseness: Yu and Bairner (2008), for instance, look at how the KMT, in the 1970s, after losing the support of the US government, looked to the wider Chinese diaspora as a means of supporting its claim to legitimate Chineseness. He (2012) considers the representation of Taiwan in Hong Kong cinema as a form of alternative Chineseness, thus associating Taiwan with the complexity of Hong Kong's own relationship with the rest of the mainland at the time of writing, as a colony of the UK until 1997, and one that has since retained a sense of having a distinct, if contested, identity. Tien and Luan (2015) base their study investigating the value of "bamboo networks" to migrants (see chapter 7) on a Taiwanese sample; since the bamboo network is said to be a particularly Chinese phenomenon, the choice of a Taiwanese subject group is significant, in that it recognizes Taiwan as having a claim to Chinese identity while also identifying Taiwanese as a group in their own right. Chiang (2011), in her study of return migration, focuses on the Taiwanese diaspora as a distinct one from the Chinese and as having a presence in a multitude of countries. Studies of the Taiwanese diaspora thus find it hard to get away from the KMT-inspired trope of the "free Chinese" identity. Finally, Tseng (2017) takes the differences between

Chinese and Taiwanese identity as the starting point for exploring transnational relationships between mixed couples.

Other studies focus on Taiwan as a key part of a global Chinese identity, one that includes but is also greater than the mainland. Lien's study of attitudes toward mainland China among diaspora Chinese in the United States compared cohorts from the mainland, from Taiwan, and from Hong Kong (2011). Zhou and Hsu (2011), similarly, compare the return migration experiences of Taiwanese and mainland Chinese people in the hi-tech sector. Ong's (1999) seminal study of transnational Chinese identities, while focused on China, recognizes that diasporic countries such as Taiwan have been instrumental in the globalization of China and the development of Asian models of globalization. The distinctively Taiwanese nature of the diaspora and the island's complicated relationship with the mainland, and the idea that Taiwan represents an alternative form of Chinese identity, are becoming of greater interest to academics, policymakers, and citizens; however, the focus is still on Taiwanese identity as a form of Chineseness rather than an identity in itself.

Studies that genuinely engage with the internal divisions and external connections of Taiwanese migrants, are, unsurprisingly, less common. Lin (2011), who looks at the migration of *waishengren* back to the mainland, is one of the few who considers the *waishengren/benshengren* distinction and its significance for the diaspora, even though that distinction has implications in terms of what languages the migrants speak, what organizations they join, and whether or not they engage with Taiwanese government–sponsored organizations (as opposed to mainland-sponsored organizations). Yang and Chang (2010) consider the *waishengren*, with some justification, as a diaspora in and of themselves. Some writers on the Taiwanese diaspora consider the blending of ethnic identity with professional and other identities: Montgomery (2008) describes the role of alumni networks for Taiwanese engineers in the United States and how they use email to maintain these connections; networks allow them to build careers and stay employed in a volatile sector. The concept of Taiwanese identity as part of a pan-Asian identity does not really seem to be considered, despite Taiwan's status as a "tiger economy" and its strong resemblances, in terms of its culture and history, to the multiply colonized nations of Southeast Asia rather than the imperial countries of China and Japan. Finally, the blurring of ethnic and national identities, always problematic, is heavily politicized in cases such as that of Taiwan.

The answer to the question "Why Taiwan?," therefore, is that to examine it more closely is to gain a greater understanding of an Asian diaspora that is widespread and gaining in influence and that has a complex and contested identity both at home and abroad, but that is seldom studied in and of itself, being instead continually considered "Chinese" in one way or another. Moreover, besides exploring the question of identity for this increasingly important country, an examination of the Taiwanese diaspora could have wider implications for our understanding of how identity in general works in constructing networks and bridging transnational social spaces, of the connections and differentiations between ethnic and national identities, and in how these are formed and continued in nationalist political environments as well as globalizing ones.

Chapter Two

The Network Society and Taiwanese Skilled Labour Migration

Before we can explore the social world of Taiwanese skilled labour migrants, we need to understand the form it takes and how people with transnational lives operate in the current period of what one might call "threatened globalization." I will explore the origins of the concept of the network society as developed in the late 1990s, the role of identity in establishing and maintaining these societies, what has happened in the two decades since the subject was first introduced, how it works in the Taiwanese case, and the threats raised to the network society by protectionist movements and by physical and economic barriers to movement.

Islands in the Net: The Network Society Concept

In order to understand how Taiwanese skilled labour migrants live and manage across borders, it is necessary to first outline the concept of the "network society." This was developed by a number of researchers in the late 1990s, but I will particularly focus on the articulation of the concept by Manuel Castells in his trilogy of social science works, *The Information Age: Economy, Society and Culture*, on how it was developed by later researchers in subsequent decades, and on the challenges that globalization in the twenty-first-century raises for researchers in this area.

Castells (1996) takes as his starting point that "our societies are increasingly structured around a bipolar opposition between the Net and the Self" (p. 3). He argues that society under globalization is "informational, global and networked" (p. 77), organized as a "space of flows" rather than as a rooted "space of places" (pp. 408–409). He describes the network society in some detail, as a society built on electronic communications, with social and economic nodes in the global cities but with raw materials, capital, and people continually in flux between them. It is not detached from physical space; rather, it exists in a complex relationship with space. For a global elite to maintain dominance, for instance, it draws on physically embedded structures of elite schools and organizations. Furthermore, local governments, paradoxically, are crucial to setting up and maintaining globalization (p. 137). Castells's society is thus organized around global cities, but flowing constantly between them.

Within this space of flows, Castells argues, industries are spread around the world like webs, and highly skilled labour migrants pursue global careers: "anyone with the capacity to generate exceptional value added in any market enjoys the chance to shop around the globe" (p. 130). Information technology is crucial to maintaining this system, allowing as it does real-time communication between places. A number of groups of skilled people, sharing a common national origin and ethnicity, maintaining connections with one another around the globe, could therefore be described as inhabiting Castells's network society.

Similar things were argued, albeit sometimes in different terms, by other writers of the late 1990s and early 2000s. Tomlinson (1999), in his book *Globalization and Culture*, argues that the relationship between global and local reflects a "complex connectivity" (pp. 2, 71). He contends that even as people engage in activities in "global spaces" – flying on airplanes, using the Internet, and other practices that cannot be said to take place in a specific locality – they are at the same time embodied and physically located (pp. 149, 141–143). Deterritorialization does not mean "the end of locality, but its transformation into a more complex cultural space" (p. 149). Similarly, Appiah's (1998) theory of "rooted cosmopolitanism" presents a world in which people are cosmopolitan in outlook and live border-crossing, globetrotting lifestyles even while retaining connections to particular locations, especially their places of origin. Vertovec (1999, 2001) generalizes global social groups under the term *transnational social formations*, meaning any group, however loosely organized or structured, that operates in two or more geographically distinct locations simultaneously (1999, p. 447; 2001, pp. 4–5). This definition covers a wide range of social activities and groups, from casual labour networks (Anderson, 2000), through to cities or societies that span a national frontier (Meinhof, 2001), through to (in more recent research) refugee communities (Bocker & Havinga, 1997; Landolt & Goldring, 2010) and criminal networks (Kshetri & Alcantara, 2015). The concept of the network society is, therefore, supported in and developed by the works of other researchers of the period (and afterwards).

One might have expected the 9/11 attacks and several other major terrorist incidents, a global recession, and rising prices for raw materials, all of which occurred in the two decades after the concept of the network society emerged, to challenge the very existence of that concept. But in fact, those events seem instead to have buttressed it. International education is no longer the province of a wealthy super-elite and for many people has become a necessary class marker (Teo, 2011). Many companies have developed an "international stream" of managers, who are expected to spend a few years in various corporate outposts before rising to senior management levels (see Forster, 2000; Hamuri & Koyuncu, 2011). Since the late 1990s, information technology has become ever more crucial to modern life: in the 1990s it was a largely the preserve of the elite, today it is far more universal. Mobile phones and wireless networks have enabled rapid communication in the developing world (Thompson, 2009; Horst, 2006), as well as complicated situations such as the simultaneous existence of heavy industry and service industries in India and China, both still technically considered developing countries (Mirchandani, 2004, Beerepoot & Lambregts, 2015). The response to rising transportation costs during the 2008 recession and to the hardening of national borders under protectionist regimes in the 2010s has largely been to substitute digital communications for physical presence, with the result that transnational capitalist activities have carried on unabated (van Veen, 2018). Globalization today is a social phenomenon engaging both dominant and countercultural groups, with labour and products permanently entangled in transnational networks.

Furthermore, since the early studies of transnational social phenomena, network societies have deepened, broadened, and become more complex as well as more socially acceptable (see Gammeltoft-Hansen & Sorensen, 2013, Gonzales, 2019). For instance, online gaming (Zhong, 2011) and what Goh (2014) calls the "global fantasy industry" have acquired forms that are inherently transnational. Another result of network society is the "born global" company (Knight & Cavusgill, 2004). Labour migration, which touches different classes and generations and has multiple causes, today engages governments as well as social institutions (Ho, 2011a, 2011b). Helbling and Teny (2014) studied cosmopolitan German elites and found that their cosmopolitanism, paradoxically, had made them even more attached than ordinary Germans to their nation (or, to be more precise, their national identity). But at the same time, there are forces working against the spread of transnational social phenomena; these include anti-migration laws, as well as barriers to non-local Internet activity such as the Chinese "great firewall," recent proposals for national or local Internets (Tajdin, 2013), and the rise of nationalist movements such as Brexit (Redclift & Rajina 2019). Glick Schiller and Salazar's (2012) overview article posits that settlement and mobility coexist in a changing and complex dynamic. The decades since the theory was first developed have seen an expansion of network societies, which are now globally dispersed and technologically embedded and take a number of innovative forms, all of them reflecting a complex relationship between motion and stasis.

A modified concept of the network society is a useful starting point for a study of Taiwanese skilled labour migrants; however, there are gaps in the research. To begin with, Castells, like many other academics of the period (e.g., Sklair, 2001; Guarnizo & Smith, 1998), focuses mainly on elite and on working-class mobility, yet other studies have indicated that there is a large middle-class cohort of skilled labour migrants who engage with globalization in their own different ways (see, for instance, Moore, 2005; Fechter, 2007; Shimoda, 2017). This middle ground includes IT professionals (Harvey, 2008), students (Teo, 2011), and managers (Moore, 2005; Beaverstock, 2018), rather than the billionaires studied by Sklair (2001) and the Dominican unskilled labourers studied by Portes (1998). Furthermore, migrants may be more (or less) globally focused depending on their stage in life – a point that Castells acknowledges, though he does not go into detail about it (Moore, 2005, ch. 7). Moreover, as already noted, research on labour migrants tends to focus on their activities in one particular node rather than considering them in the context of the space of flows. Also, there is little acknowledgement of the differences that location can make: most of the above-cited writers treat global cities largely as if they were all basically the same, and similarly treat labour networks as if (barring certain superficial differences) they are all more or less the same phenomenon (see Moore, 2017). When studies of networks focus on a single characteristic (ethnicity, for instance, or shared university education, or obligations such as the Chinese practice of *guanxi*), they can obscure the complex connections between forms of networking that happen in practice. Finally, the complex interactions between global and local discourses in this space of flows are still not well understood; for instance, Jiang and Fung's (2019) recent study of online gaming indicates that national and regional discourses combine with global ones in complicated ways, and some studies of nationalist movements indicate that, despite their ostensible rejection of globalization, the on-the-ground reality is more complex (Redclift & Rajina, 2019).

So, there is room to expand the basic concept of the network society, not just to take into account the impact of recent technological, political, and cultural developments, as with Rehnberg and Ponte (2018) or van Veen (2018), but to actively consider the experience of skilled labour migrants in the space of flows, and the ways in which the locations and networks they connect with affect their experiences. The potential of the network society concept as a means of understanding the world of skilled labour migrants has thus not been fully explored in the two decades since its formulation. While Castells's theory still broadly holds, it faces challenges from the more complex situation outlined by Glick Schiller and Salazar (2012) and by others; it needs to be developed to consider the complex and contradictory reality of the contemporary "space of flows."

It's Who You Are: The Role of Identity in Network Societies

A crucial point emphasized by the theorists noted earlier is that the glue holding these networks together is identity. As Castells (1996) puts it, "identity is becoming the main, and sometimes the only, source of meaning in an historical period characterised by widespread destructuring of organizations, delegitimation of institutions, fading away of major social movements, and ephemeral cultural expressions" (p. 3). Identity, furthermore, is flexible and multivalent, such that a national identity might refer to a connection with a particular territory or ethnic group without requiring the holder to actually be in that location or even be a member of that ethnic group (see Moore, 2005), making it an ideal way for a network society to maintain cohesion. The complicated and contested nature of Taiwanese identity will, therefore, be crucial to understanding Taiwanese skilled labour migrants as a network society.

At this point, it is worth focusing on Castells in particular, as the second volume of his trilogy (1997/2004) addresses in some detail the importance of identity to the network society. He begins by arguing that "in the network society ... for most social actors, meaning is organised around a primary identity ... which is self-sustaining across time and space" (p. 7). This does not preclude the holding of other identities, but, he argues, usually one particular identity dominates. He surmises that this is because of the disintegration or transformation of traditional institutions and nation-states (pp. 11, 420), and in this regard, he devotes a section to the role that identity played in the break-up of the Soviet Union (pp. 35ff), adding, however, that this was also due to the increased deterritorialization that is a prominent feature of the network society. Like the network society itself, however, identity is not detached from location; rather, it is in a complex relationship with often multiple locations as well as the spaces between them: Castells points out the paradox that "the age of globalisation is also the age of nationalist resurgence," noting that the latter takes the form of both "the challenge to established nation-states and the widespread (re)construction of identity on the basis of nationality" (p. 30), with, again, the question of the relationship between ethnic and national identity being heavily problematized. Finally, Castells notes that such projects are embedded in power relations: "who constructs collective identity, and for what, largely determines the symbolic content of this identity, and its meaning" (p. 7). So if Castells's theory of the network society applies in this case, it is crucial to look at how identity is used, especially in cases where ethnic and national identities are conflated within nationalizing projects or reconstructed as part of nation-challenging ones.

It is noteworthy that Castells's concept of identity, like that of many writers on identity in transnational contexts, draws tacitly on the concept of identity as performance, as developed by Erving Goffman in the mid-twentieth century (1959, 1961, 1963, 1967) and somewhat revived around the turn of the millennium (Moore, 2005; Spence, 1998). This approach focuses less on the psychological side of identity and more on identity as a kind of performative discourse whereby individuals and groups display identity through symbols. Furthermore, while Goffman's original theory did not consider this, later writers have focused on identity as developing through a kind of nuanced dialogue in which, at the same time as we present ourselves strategically, our self-presentation is being interpreted by, and incorporated into, the self-presentations of others (Jenkins, 1996, p. 58; Cook-Gumperz, 1983, p. 123; Cohen, 1994, p. 12). This leads to models such as Brannen and Salk's (2000) "negotiated culture," in which self-presentation is seen as a form of negotiation between different parties, with various aspects being accepted, rejected, and agreed on in a continuous process. Later writers have added the dimension of power and resistance to this theory, noting that whose portrayal of identity dominates is far from neutral (Jenkins, 1996). Identity, as self-presentation, is therefore not a fait accompli or a performance that the audience must passively accept, but an active dialogue between members of the group and outsiders – and, indeed, within groups themselves (see Alvesson, 2002; Alvesson, Ashcraft, & Thomas, 2008).

Finally, the concept of identity in the network society is also predicated on the idea that identity is expressed through symbols. Symbols, it has been noted, are flexible and "multivalent," that is, they can simultaneously hold multiple meanings for their perceiver (Sperber, 1974). As such, they are often used in ritual and political contexts, where their meaning becomes the subject of power relations as various parties attempt to define and thereby control the meaning of the symbols (see Bloch, 1974; Strecker, 1988). The role of symbols in defining communities and policing their boundaries has often been noted for small-scale, non-transnational societies, as in Anthony P. Cohen's studies of how symbols of identity were used in a small Scottish island community (1985, 1986). However, they have since proved, if anything, even more important for the construction of transnational societies, which, without grounding in physical location or other concrete objects, increasingly focus on symbols as sources of identity.

A number of contemporary and subsequent studies have supported Castells's formulation of identity in the network society. In Baumann's contemporary account of the multiethnic English suburb of Southall, for instance, both official doctrine and direct questioning of inhabitants suggests that the neighbourhood is comprised of unified ethnic groups that express their identity through collectively held symbols that define the groups' boundaries. However, Baumann's (1996) observations of people's expressions of group allegiance suggest that these symbols come into play, not so much in defining boundaries between ethnic groups, but in communications between individuals as they negotiate the interweaving of their various frames of reference. Furthermore, symbolic self-presentation can be continually altered to fit the social context, a crucial property in transnational social interactions. Banks (1998) speaks of the cross-culturally variable linkage of form and meaning with regard to images: a particular photomontage is not read the same way in India as in the UK, and both it and its readings must be considered in context.

This has also been shown in the case of more recent anthropological and geographical studies of skilled labour migrants, such as the Taiwanese migrants who constitute the core of this study. Beaverstock, for instance, explores how British expatriates in Singapore use their ethnic networks to further their careers (2011), and Ryan and Mulholland (2014) note similar practices among French expatriates in London. The symbolic nature of self-presentation also emerges in the paradoxical results of Helbling and Teny (2014), whereby a national identity can be used to support group cohesion in cosmopolitan social spaces. The fact that self-presentation is conducted through the expression of multivalent symbols means that it can be part of the strategies of skilled labour migrants in their interactions with one another (Robertson, 1992, p. 166; Burns, 1992, p. 232). For transnational elites, the strategic presentation of identity is a means of maintaining social connections while transcending physical borders.

In the case of the Taiwanese skilled labour migrants, therefore, it stands to reason that identity is a particularly important way of maintaining a consistent presence throughout the space of flows. However, the Taiwanese migrants present a unique way of exploring how a contested identity might operate in this context. The same symbols of ethnic identity are used by both *waishengren* and *benshengren*, albeit with different meanings and different relationships to the concept of Taiwan as a nation; the Taiwanese also use symbols of Chineseness common to the entire Chinese diaspora, but in ways that subtly indicate a distinctive identity. Finally, the migrants must do this in different social contexts, as well as in the transnational social spaces they negotiate in order to maintain themselves as a transnational network of skilled labour migrants. Besides being an identity that has historically received little attention, therefore, Taiwanese identity provides an interesting expansion and development of Castells's original theories about network societies and the power of identity, and a way to understand what has happened to globalization in the meantime.

Uneven Globalization and the Taiwanese Diaspora

A study of Taiwanese skilled labour migrants, finally, provides a way of developing earlier studies of identity theories by highlighting the complex relationship between local and global that is a key aspect of a society that exists in a space of flows punctuated by nodal points. "Unevenness" has long been acknowledged as an aspect of globalization, with different locations and groups having different degrees and kinds of cross-border social ties. Only now are researchers beginning to study unevenness as a phenomenon in and of itself, and the effect it has on transnational social formation. For instance, while the differences between global cities have long been acknowledged, little work has been done on how this affects the engagement of individuals living in those cities (Moore, 2016), and the same is true of many other groups and locations in global networks.

Furthermore, many studies tend to operate on the tacit or explicit assumption that "globalization" and "transnationalism" are essentially a single phenomenon and that all activities that cross borders share certain traits. Castells does consider different manifestations of the network society, from labour migrants to nationalist movements in former Soviet republics; he focuses on their common points rather than on the differences that might stem from variations in ethnicity, origin, class, and so forth, and how these affect the ability to engage with the space of flows. This gap is significant, because these differences are likely to affect the opportunities enjoyed by individuals in the home and host environments, the ability of businesses to operate successfully across borders, and the nature and extent of transnational connections that can flourish. As Holý (1987) noted in the introduction to his classic work on the comparative method in anthropology, the search for contrast is inherently complementary to the search for similarity and can bring out perspectives and phenomena that a search for similarities cannot. The unevennesses of globalization requires more exploration, from both a research and a practical perspective, in order to understand what it really means.

Finally, multisited studies of the same group, while not unprecedented (see Hage, 2002; Falzon, 2016) are still relatively rare. This is particularly the case regarding the Chinese diaspora; there are many studies of return migration, for instance (Teo, 2011; Zhou & Hsu, 2011; Tsang et al., 2003), which have an implicit multisited aspect, but these tend to look solely at migrants once they have returned rather than considering them first in one site and then in the other. A study of different nodes of a transnational ethnic network, looking at the differences that location, network, connections to the diaspora, and professional activity make, and comparing and contrasting how this occurs in different areas and across a variety of networks, can be used to develop a more nuanced image of globalization, global cities, and transnational networks, one that takes both similarities and contrasts into account.

Studying the Taiwanese skilled labour migrants will allow an analysis of a network society in terms of the impact of different locations, of different ethnic/diasporic networks, and of different social networking organizations, and the differences and similarities these make to the opportunities, local and global engagement, and transnational connections of a single, ethnically focused migrant group. The different nodes in the Taiwanese section of the network society will be considered not just at different locations, but using different lenses: considering them as the residents of global cities, as people participating in discourses of Taiwanese identity, and as members of different sorts of transnational social networks.

What Follows After: The Taiwanese Migrant Study

The rest of this book will focus on three cohorts of Taiwanese skilled labour migrants, at three nodal points in the wider Taiwanese global network. The data were gathered by a single researcher (an ethnically European cisgendered woman with a working knowledge of Mandarin, born in Toronto and living full-time in London, aged between thirty-five and forty at the time of the study) in three physical locations (the UK, Canada, and Taiwan) and also online: via email, on forums, and on social media such as Facebook and a now-defunct Scottish-focused local social media site named KILTR. Forty-eight individuals were formally interviewed (see Appendix 1), and participant-observation was carried out over five years at community social and business networking events in and around the Taiwanese communities of London and Toronto. Social and cultural background research was conducted in Taipei and Kaohsiung.

Three cohorts form the core of this study: the British group, consisting chiefly of networks around the Taipei Representative Office and the European Taiwanese Chamber of Commerce; the Canadian group, developed through the networks around the NTU Alumni Association and personal and professional social ties; and the Taiwanese group, which involved the alumni, faculty, and students of National Chengchi University. These communities were all connected to three global cities: London, Toronto, and Taipei; however, these cities were the *focal points* of the different Taiwanese communities' activities, rather than their exclusive locations. For instance, while the cohort interviewed for the UK portion centred on London, some of the people involved were based in Telford, Manchester, Oxford, or other cities. Likewise, the Canadian network extended out of Toronto into Vancouver and the United States, and the Taipei component of the study included forays out of the city and a visit to Kaohsiung. But because these cities did form the social and geographical focus of each community, they will be referred to in the material that follows as the "Toronto cohort," the "London cohort," and the "Taipei cohort." Note here that this book does not consider the position of the indigenous people of Taiwan, even though they are undoubtedly important, because I did not encounter any informants who self-identified as indigenous Taiwanese, so my findings must not be taken as in any way reflective of their position (which has in any case been extensively studied, for more on which, see, among others, Brown, 2001). I will now introduce and describe the different cohorts, the ways in which they defined and explored their identities as Taiwanese, and the conflicting discourses that surrounded them.

A Brief Note on Methodology

Interviews

The interview portion of the study consists of interviews with forty-eight individuals in, or associated with, London, Taipei, and Toronto (see Appendix 1). The characteristics of each group will be discussed in more detail in the chapters relating to each cohort and city. I will note here, however, that the interviews were generally between forty-five and ninety minutes in length and were semi-structured, in that I developed a few questions beforehand but otherwise let the interviewee lead the conversation. Where necessary, some interviewees were reinterviewed, either to answer questions that had arisen during the interview or to follow up on an incomplete interview. These data were also supplemented with informal conversations with participants outside an interview setting, and in some cases online. While sampling will be discussed more thoroughly in the chapters relating to each cohort, broadly speaking I was looking for individuals with a university education or higher either working or seeking to work in a professional field. All of the interviewees self-identified as "Taiwanese," with some also self-identifying as "Chinese."

Interviews were conducted mainly in English, with some use of Mandarin for clarification or where English did not lend itself to explaining the concept under discussion. In the Taipei portion of the study, a translator was usually present in cases where my level of Mandarin was not up to the complexities of the interview, and two interviews were, in fact, conducted mainly in Mandarin by the translator acting at my direction, with some deviations into English. The interviews were transcribed by a (non-Mandarin-speaking) professional transcriber, with corrections by the researcher and by Mandarin-speaking research assistants as needed. Pseudonyms are used for all interviewees quoted in this book, and some identifying details have been changed. As far as I am aware, all interviewees and assistants were ethnically Han, except for two Hakka, who identified themselves as such during the interview.

Participant Observation

Participant-observation was conducted at networking and social events for the Chinese and Taiwanese communities in both London and Toronto. In London, these included the biannual Taiwanese Food Festival, the Lunar New Year festival, and meetings of the European Taiwanese Chamber of Commerce, as well as language schools and language learning groups. In Toronto, these included a Taiwanese night market and visits to local community sites popular with the Taiwanese, such as the Pacific Mall, the Metro Mall, and Downtown Toronto Chinatown. During the visiting fellowship in Taipei, I also conducted ethnography in and around the Taipei area, through living in the city, working at NCCU, and travelling to cultural sites of interest. The Taipei visit in particular was useful in terms of providing cultural background for the study and providing a general understanding of the starting point the Taiwanese overseas were coming from, as well as developing an understanding of the *habitus* (Mauss, 1934) and tacit symbolism of a particular culture through the lived experience of observing and participating in it.

Online Networking

Some data were also obtained through online participant-observation, or "netnography" (Kozinets, 2009), in line with the growing number of studies considering the extensive role that digital technologies currently play in the networking of migrants, skilled and otherwise, and the development of global social and economic connections (see, among many others, Dekker & Engberson, 2014; Rehnberg & Ponte, 2018; Gonzales, 2019; Alinejad et al., 2018; Marlowe, 2019). This was achieved through joining online communities, forums, and Facebook groups aimed at Taiwanese community network-building, as well as communicating with Taiwanese friends, colleagues, and informants in London, Taipei, and Toronto. In Taipei, also, blogging and microblogging about my experiences allowed me to benefit from the insights of friends and colleagues overseas. These online activities allowed me to observe the ways in which new information technologies are being used to build connections, advertise events, recruit volunteers, and express particular social identities, in a setting that allows and even encourages transnational interaction. I also conducted some interviews via email, particularly with individuals in North America and elsewhere, owing to the impossibility of arranging face-to-face interviews and the difficulty, due to time zone differences, of arranging telephone or Skype interviews. Again, however, as electronic media are often used by transnational professionals to maintain contacts and develop transnational networks, this experience added an extra layer to the ethnographic aspect of the study.

Analysis

The resulting data were coded by reading the interview transcripts, participant-observation reports, and surveys closely, then highlighting relevant extracts in different colours, each colour assigned to a different broad category (see Brannen, Moore, & Mughan, 2013). These categories were then broken down further into subcategories. Excerpts were chosen either to represent a general opinion or when a particular observation was unusual or insightful in some way. The analyst also reflected on their own participation in the research and analysis process, with a view to remaining aware of the role their own identity played in the process, and kept up a dialogue with some interviewees (who had agreed to continued participation in the project) about the results and their interpretation. It should be noted that this analysis does not make any claims to represent a general "Taiwanese perspective." Indeed, the diversity of such a perspective is one conclusion that should be extracted from the project's findings.

A Note on Suitability and Reflexivity

That I was able to study three groups in different locations provided insights into the impact of location. Having London and Toronto as the locations of the two migrant communities was particularly useful because, first, it allowed a comparison of European and North American migrant experiences. Second, although London is often said to be one of the three most "global" cities in the world, the others being Tokyo and New York (Sassen, 2001), and is often studied from that perspective (e.g., Ho, 2009; Moore, 2005; Leyshon & Thrift, 1997), Toronto's global nature is far less frequently considered in detail, and the comparison may raise wider questions of what it is for a city to be "global." Finally, the presence of the Taipei cohort allowed for some comparisons with the same group's experiences "at home" and with the symbolic, emotional, and social starting point from which the diaspora had emerged.

Furthermore, the fact that the Taiwanese groups in question centred on a small number of business and alumni associations had a number of benefits. First, it provided a means of studying how these organizations facilitate transnational networking, following on research in this area by Cordero- Guzmán (2005), Schrover and Vermeulen (2005), Beaverstock (2005, 2011, 2018) Beaverstock and Smith (1996), and others. Second, it organically followed the processes by which transnational professionals construct cross-border networks. This study offers the potential for understanding not only the factors affecting the growth and development of transnational networks and the ways in which migrants adapt to their local communities while maintaining a distinctive presence, but also the role of network-building organizations, ethnic diasporas, and location-bound traits in these processes.

Finally, while the use of the "snowball method" for obtaining interviews might raise the argument that the sample is not demographically representative, I would note that the point of the study, and of the use of a primarily ethnographic method, was not to identify the "typical Taiwanese" experience, but to understand the lived experiences of individuals in building their networks. As the "snowball method," which involves asking one's extant contacts to recommend further interviewees, replicates the process of network-building, it allows insights into the process of developing transnational networks.

With regard to the researcher's characteristics, I am a Toronto native who was educated at Oxford between 1997 and 2002 and has been living in London since 2003. I am also a professional academic in the area of international business studies and human resource management. Like my interviewees, I could be described as a transnational professional, or skilled labour migrant, with connections to the locations in the study. So I have used my personal, professional, and family networks, and shared personal experiences with the study's participants, to develop contacts and learn about the Taiwanese communities in both cities. This process, essentially, follows the same one that the Taiwanese people who form the subject of the study use when developing contacts in their birth and adopted countries, and, later, through their professional networks. Furthermore, as noted above, the use of my pre-existing social and professional ties to launch and continue the study means that the process of network development was ethnographically replicated for study purposes through the research process.

However, it must also be noted that my connections with all three sites affected my role and relationship with my interviewees, the sort of data that were shared during the interview process and that I obtained through participant-observation, how I was perceived, and the role to which I was assigned by the study's participants. As an academic with business connections, my network inevitably became skewed toward researchers and businesspeople. Given the relationships and rivalries between UK universities and the role of alumni networks (see chapter 8), it may also be significant that the student representatives from the University of Oxford were willing to provide me with interviews, but not those from the University of Cambridge.

Everybody involved directly with the study was familiar with the concept and value of social science research, so I did not face the problem of having to explain my activities to people for whom the concept was strange and suspicious (unlike, for instance, Briggs, 1986). I did, though, have to deal with the fact that people had preconceptions as to what social science research might involve (a common problem when conducting anthropology among urban or elite groups; see, for example, Fox, 1977), and so could be surprised when they found that my data collection methods did not include gathering and analysing statistics or writing formal reports for public policy-making bodies. However, bearing in mind these facts as part of the reflexive character of the research, the data that follow can be seen as a particular image viewed through a specific ethnographic lens in the context of the UK, Asia, and Canada in the early twenty-first century.

This study of Taiwanese skilled labour migrants is, therefore, not just a way of understanding a new global power with a particularly contested national identity. It also explores an aspect of the network society that has not been fully explored by subsequent researchers in this area, by analysing the impact of different locations, networks, and social organizations on the experiences of transnational labour migrants, and uniting the two by considering the role of identity – in particular, contested identity – in global contexts.

At a time when globalization is in flux, challenged by the rise of protectionist movements yet encouraged by the increasingly networked nature of business, production, political activity, and even crime, it is crucial to understand how groups like the Taiwanese use their identities to construct and maintain a network society with a space of flows.

Chapter Three

Signs and Meanings: Defining and Maintaining Taiwanese Identity

Having introduced the concept of identity as a performance enacted through the use of symbols, it is worth briefly discussing, comparing, and contrasting the common symbols of Taiwanese identity encountered in the different locations before we look at these groups in more detail. Note that this list is not comprehensive and that other symbols (as well as other contestations and discourses surrounding them) will emerge later in the book. I will focus here on symbols related to geographic and/or genetic connections to the island of Taiwan: food, holidays, and language.

Island Connections

A connection with a particular physical or geographical region is usually cited as key to national identity (Smith, 1995). Some writers (e.g., Hofstede, 1980) extend this to associating particular traits, or combinations of traits, with specific nationalities. For instance, among German transnational business people, two characteristics that emerged as defining their ethnic identity were *Blut* and *Boden*, "blood and ground" (Forsythe, 1989, Amiraux, 1997). In other words, to be German, one had to have been born in Germany or of German descent (Moore, 2005). In practice, however, these traits are considerably more complicated. Claims by the descendants of Turkish, Italian, Korean, and other "guest workers" to Germanness on the basis of having been born in Germany are frequently cited as examples of contested identity (White, 1997). By contrast, within the German diaspora, "blood" alone has not always been necessary or sufficient grounds for claiming a German identity (Forsythe, 1989, p. 153).

For the Taiwanese, these concepts can be even more complicated. Taiwan is a colonial country, as noted above, so claims to Taiwanese "blood," outside of the indigenous community, are problematic. At the same time, connections to Taiwanese "ground" are also open to contestation, for the *waishengren* were not born in Taiwan and their descendants may view themselves as Chinese rather than Taiwanese. Indeed, until quite recently, *benshengren* were encouraged to see themselves as such (see Hall, 2003; Liu & Hung, 2002; Chen, 2012). While this will be discussed in more detail later, it is worth noting here that one of the KMT-affiliated participants in my study was born in Singapore, lived in the UK, and identified as Taiwanese purely as a political act. Interviewees often mentioned Taiwan's history as a seafaring nation, suggesting an identity also built on migration. Although a geographical connection with the island did emerge from people's self-identification, it was not as superficially straightforward as it is for many other ethnic groups.

Another influence in this regard is the idea of Taiwan as a "free Chinese" or "pure Chinese" culture (Hall, 2003). Museums and cultural centres in Taipei often present history as if the mainland ceased to exist in 1949: the Imperial Palace Museum, for instance, displays historical artefacts from mainland China up until the 1940s, at which point its scope contracts to the island without explanation, as if there is a natural continuity between present-day Taiwan and the earlier history of China. More locally focused museums have similar takes on history, acknowledging Taiwan's shifts in colonial possession more directly than the Imperial Palace Museum, but with the Japanese occupation presented ambivalently or in contradictory terms (a Confucian temple and the Lin Liu-Hsin Puppetry Theatre Museum both position Japanese rule as oppressive, whereas the local museum in Xinbeitou, a spa town near Taipei where the Japanese established bath houses, is more positive). Taiwan's history of multiple colonizations may explain why a national identity associated with a particular geographic region or genetic heritage is less prominent, and indeed more complicated.

Returning to the German comparison, it is also worth considering Taiwan's economic "miracle" of the 1980s, when the island rapidly developed its electronics industry and went from a producer of cheap electronic goods to the home of prestigious global brands such as Acer and HTC (Hall, 2003, p. 154). In the case of the Germans, during the postwar years, national identity expressed as *Blut* and *Boden* was tainted by association with the Nazis, so the focus of that identity shifted to corporations of German origin, which served to channel German identity into something associated with the nation rather than the Nazis (Head, 1992; Weidenfeld & Korte, 1991, pp. 149, 154). While Taiwanese people did sometimes speak of national brands with pride (for instance, joking that as a Taiwanese, one had to own an HTC phone rather than one of the more common Samsungs, out of national loyalty), they did not seem to have anywhere near as strong an allegiance to their companies as the Germans cultivated (or the Canadians, regarding Blackberry).

Food

Food, both specific local dishes and a more general cosmopolitan fondness for different foods, was important in all three locations in this study, but in significantly different ways. The annual Taiwanese Food Festival was a crucial event for the London community, and also central to that community was a small business that specialized in importing Taiwanese food products (and that was present at every community event I attended). In Toronto, one identity marker was a local chain of pan-Asian grocery stores, T&T;[1] those stores, which had been founded by an ethnically Taiwanese couple whose daughter had subsequently taken over the business (Ebner, 2009), sponsored a Night Market on the waterfront, based on the Taiwanese custom of after-hours markets for street food; it was advertised prominently, though not exclusively, in Taiwanese venues around Toronto.

Food, as a symbol, branched out in other directions. The London event was advertised and written up on more generally aimed food blogs with the intention of attracting London's "foodie" community, that is, cosmopolitan Londoners interested in exotic gourmet foods; this festival was led by student associations and attracted non-Taiwanese visitors and participants. In Toronto, the T&T stores are pan-Asian rather than exclusively Taiwanese or even Chinese, and cater not only to the Asian community but also to non-Asians looking for exotic food items (or even, since the stores also stock a standard range of grocery products, simply shopping). In terms of how the wider community constructs Taiwaneseness, it is interesting that a pseudonymous reviewer for the popular Toronto blog blogto.com describes the T&T stores in this way: "all of the dingy stores in Chinatown bow down to its shiny, antiseptic aisles and professional service" (www.blogto.com/city/2007/08/tt_grocer_mastery), implying an identity that is Chinese but also clean and modern, in contrast to that of Chinatown (the blog implies Downtown Chinatown, more associated with mainland Chinese and Hong Kong Chinese migration), which is seen as old and dingy. Similarly, although the night market may be based on a Taiwanese custom, it offers foods from all over the world, and lot of the ostensibly Asian food is fusion cuisine – that is, it offers Asian dishes with

1 For some informal background on this chain, see www.blogto.com/city/2007/08/tt_grocer_mastery.

new ingredients or non-Asian dishes with Asian ingredients. Like the London food festival, it attracts many people without a Taiwanese connection.

In Taipei, in contrast to the diasporic sites, local food was less of an identity marker; there one found instead a cosmopolitan taste for different sorts of food. People were constantly discussing, and sharing information about, foreign or exotic cuisines such as Japanese, Italian, German, rural mainland Chinese, and so forth; one interviewee, a Canadian former expatriate, told me that Italian restaurateurs had a significant presence in Taipei. Another ambiguity: when I remarked on the Taiwanese interest in food, many people responded by telling me they thought it was a general Chinese trait, making jokes about Chinese people "eating anything that doesn't eat them first." This cosmopolitanism was to some extent present in the diaspora. One of the first indications, for me, that the Metro Mall in Toronto was a focus of Taiwanese identity, and not Chinese identity more generally, was that its food court included a very cosmopolitan range of restaurants – albeit less obviously than in Taipei; people were less interested in food generally than in food from the Asian region or fusion cuisine. While food was a significant identity marker at all times, there were shifts in terms of what that denoted, as well as connections with other identities linked to this set of symbols.

Holidays and Celebrations

Several Buddhist and one evangelical Christian organization were often present at Taiwanese community events in London, yet religion did not seem to be a prominent marker of identity there, with some notable exceptions, largely related to particular life stages or rites of passage. Older informants, for instance, were often more involved with such organizations; and while I was conducting participant-observation at a Taiwanese Buddhist temple in central London on a weekday, I noticed several young couples paying brief visits to the shrine – presumably, this had to do with marriage or fertility. This is significant, because one identity point in Taipei is that Taoism is practised on the island in traditional ways that are no longer encouraged on the mainland. In London, though, Taoism does not feature significantly as an identity marker, and Buddhism and Christianity only sometimes. In London, I also observed a memorial ceremony for Dr Sun Yat-sen, carried out under the auspices of the Taipei Representative Office; but this seemed to be more a political gathering than a rallying point for ethnic identity, given that most of the attendees I spoke with appeared to be official representatives of Taiwanese political or social organizations or prominent community figures.

Both diaspora communities *did* focus on certain holidays, particularly the Lunar New Year and the Mid-Autumn Festival. (I was unable to observe either event in Taipei, but I did in both London and Toronto.) In London at least, it seemed that the Taiwanese community held its own separate celebrations for the Lunar New Year and Mid-Autumn Festival, even though these were shared events with the mainland Chinese and other Chinese groups. Nonetheless, these festivals expressed ties with mainland China and with other areas of Southeast Asia where these holidays are celebrated. It was also certainly the case that some people would have attended the wider celebrations in Chinatown, or sponsored concerts and events, instead of (or as well as) the Taiwanese ones. At the official, Taipei Representative Office–sponsored celebration in London, a mainland dance troupe had sent members to promote its upcoming tour of Europe. Food is a major aspect of most holidays, meaning there was some convergence of symbols at these festivals. Holidays celebrations were thus key performances of Taiwanese identity, albeit in different ways in different places.

Language and Characters

Language was another crucial but complicated symbol. Several interviewees from different cohorts emphasized that the use of Mandarin was distinctively Taiwanese. The use of traditional (rather than simplified) characters was especially emphasized as a source of identity. One professional language teacher in London noted: "I believe that full-form [traditional] Chinese characters will contain all the Chinese culture, history, background so [at] my school I teach the full-form … Because I studied Chinese literature, and we learned the Chinese characters because each characters they contain Chinese history, background or the people's living, all in a square picture of that character" (Hua, language teacher, 60s). Note that in Hua's justification for focusing on traditional characters, we see not just a point of Taiwanese identity, but another reflection of the "Free Chinese" identity, suggesting that the characters that form a part of Taiwan's distinctive identity are also connected inherently to Chinese history and culture in a way that the simplified characters more common on the mainland are not. In Taipei, Mandarin written with traditional characters was the most common form of language that I could see, and, furthermore, there was a tendency to use the older Wade-Giles romanization system rather than the pinyin system favoured on the mainland (see Wang, 2004).

However, the literature suggested that the use of Mandarin was, in fact, a contested symbol of identity. Although in my study, *benshengren* as well as *waishengren* seemed to take it as a Taiwanese identity marker, Hall (2003, p. 149) notes that prior to 1949, the main dialects spoken in Taiwan were Holo and Hakka, and that after that year, Mandarin was largely imposed by the KMT as a *lingua franca*. One *benshengren* informant also used Japanese as a kind of internal identity marker, talking about how her parents' generation had been educated in Japanese; this underscored the country's past history as a Japanese colony, as well as *benshengren* nostalgia for those years. Simplified characters were also to be seen in Taipei (much to my relief as a non-native speaker), and it seemed to be tacitly acknowledged that the convenience of simplified characters sometimes carried more weight than concerns about identity. Language use, and characters, could be used to challenge symbols of Taiwanese identity, as well as to affirm that identity, and sometimes even do both at the same time.

The use of Mandarin and traditional characters as a distinctive form of spoken and written language also provided connections with the wider Chinese diaspora. Some groups in that diaspora favour traditional characters, and some employ Wade-Giles romanization; my interviewees described these preferences as Chinese but as different from what was found on the mainland. Such groups were also seen as having ties to the Taiwanese; the Taiwanese social events in London and Toronto often drew people from the Malaysian Chinese and Hong Kong Chinese communities, and one younger interviewee in London talked about how she found it easier to socialize with European-born Chinese people than with mainland Chinese. London also had language clubs, which had formed around the speaking and practising of Mandarin; while the club I attended had a Taiwanese core group, it included not only other Asians but also some Europeans and North Americans who had learned Mandarin in Taiwan or elsewhere and who wanted to practise their skills. At one point a Taiwanese member showed off his skills in Holo, describing it to me as "the real, original Taiwanese language." Language was thus a complex and flexible symbol, with connections with other groups and internal contestations.

A brief consideration of some of the more common symbols and social events used in defining Taiwanese identity, together with the variations these take across borders, provides insights into how the network society defines itself through collective identity, expressed in different contexts. As outlined by Castells (1997/2004), the symbols discussed here serve as a means to develop collective identity. Yet at the same time, they are a medium through which to contest, challenge, and change that collective identity and to assert local distinctiveness.

Chapter Four

London: The City of Sojourners

In the next three chapters, I trace the journey of the Taiwanese skilled labour migrant study, looking at each cohort in turn and considering the characteristics of each node in the network, before exploring the differences these characteristics have made for the migrants and their experiences. As the study began in London, we will begin with the UK-based cohort.

The First Fieldsite: Background and Context

Origins of the Study

The project as a whole began in 2009, initially as a Nuffield Foundation–funded study of the use of ethnic identity by Chinese professionals in London to develop transnational networks and conduct business across borders. The idea was to conduct a loose follow-up to my 2000 study of German finance workers in the City of London and their uses of ethnic identity and strategic self-presentation (the results of which are summarized in Moore, 2005, and in many subsequent publications). However, although I joined mailing lists aimed at Chinese people in the UK, and was able to get interviews with a small number of Chinese professionals and representatives of community organizations, it proved difficult to identify and gain access to Chinese business networks.

I shifted my focus to the ethnic identity of Taiwanese skilled labour migrants after being introduced, by a Taiwanese colleague, to a representative of a Taiwanese business association, who opened access for me to a Taiwanese business network. Through this contact, I was able to identify and tap into ethnic identity-based networks of the kind I had been seeking for the study; via the "snowball method," I was able to gain access to the communities that had formed around, first, Taiwanese chambers of commerce and other trade organizations; second, Taiwanese student and alumni associations; and, third, the biannual Taiwan Food Festival, which linked the two (in that it was run by student and community groups, with participation and support from, among others, local Taiwanese-owned businesses and business networking organizations).

This introduction to the Taiwanese community was serendipitous, and it proved particularly useful from the point of view of identity research. As discussed in chapter 1, the complexities of Taiwanese identity – an ethnic identity on its own that also participates in a wider Chinese ethnic identity (Chow, 2012; Danielsen, 2012) – further complicated by the existence in the UK of both a Taiwanese and a wider Chinese diaspora (which incorporates other Chinese diasporas), provided an opportunity to examine a complex and multivalent identity in a transnational context, rather than a single, solidary ethnic community. By researching Taiwanese identity in the British context, and exploring it through ethnographic research, I hoped to reveal new aspects of life in the network society.

The connections I developed for the London portion of the study were based on my status as a skilled labour migrant, Canadian by birth and British by citizenship, who lives in the London area and works at a university that is part of the University of London group. The fact that I teach in a School of Management allowed for connections to the local business community through trade associations and through the school's own links to Taiwanese businesses. Contacts were obtained through university connections, through the above-mentioned trade associations, and through participation in the Taiwanese Food Festival (where I would contact representatives of businesses and organizations that were exhibiting wares or selling food). The social group referred to in this text as "Xinbeitou 8" maintained a presence at community events, and after talking with their representatives on these occasions, I wound up attending a few of their meetings, in the process also gaining access to other Taiwan-focused Facebook groups and online networks. I was thus able to use my personal connections in London, and my experiences of academic and professional life in Oxford and London, to develop the British portion of the study.

Finally, further London-centred research has been ongoing since the formal conclusion of this study: I have maintained friendly social ties with some of the study's original participants and have been introduced to other interviewees through professional contacts. I have also continued the project through academic discussions at work and at conferences and have stayed in touch with several of the organizations involved, particularly via online groups.

The London Cohort

In London, formal interviews were conducted with twenty-six individuals. Most of them were involved in one or more networking organizations aimed either at the Taiwanese community specifically or at the ethnic Chinese diaspora more generally, such as a student organization or chamber of commerce; a smaller number were associated, directly or indirectly, with a UK university and thus formed part of my own professional network. All were either professionals, retired professionals, civil servants, or postgraduate students. The professionals and students were mainly in advanced business services (law and finance), medicine, or entrepreneurship; two, however, were full-time NGO employees, and two others were university professors in business-related fields. Eighteen of them self-identified as "Taiwanese"; what this meant to them would be explored in more detail in the interviews. The rest identified as diaspora Chinese in the London context and were the founders or coordinators of diaspora Chinese organizations in London that were potentially or actually used by Taiwanese business people. Some subjects were interviewed twice and/or provided me with information through informal conversations at networking events or through email and other electronic exchanges. Three interviews were conducted with two informants rather than just one. Also, informal conversations were held with around ten additional participants at various networking events; these covered similar ground to the formal interviews. I will now give brief biographical details of one older, and one younger, member of the London portion of the study, by way of highlighting the experiences of this cohort.

The Entrepreneur: Maria

Maria had come to the UK to study physiotherapy, encouraged by British lecturers in Taiwan. While there she met her husband, who is also Taiwanese, and both settled in London. She left clinical practice when her first child was born. While living as a housewife and stay-at-home mother, Maria joined a number of Taiwanese community organizations. When a small, Taiwanese-owned travel agent that arranged travel for most of these organizations closed after the death of its owner, Maria and her husband started a replacement travel business, specializing in travel to and from Taiwan, but also China, Thailand, and other parts of East Asia. At the time of the study, the agency was a successful small business, preferred by Taiwanese business organizations but also getting a lot of business from Taiwanese prospective students visiting British universities and

Taiwanese universities arranging trips to the UK. Maria remains an active member of many Taiwanese community and business organizations.

The International Professional: Wendy

Wendy had come to the UK to study political science at the University of Manchester. While there, she had become active in the university's Taiwanese Student Association, and afterwards she remained active at local and national levels in Taiwanese student and alumni social media groups. Afterwards she was offered a job in Brussels, but she was unable to obtain a work visa and wound up taking a job in London with a human rights NGO. She was engaged to an English man when we first met, and they subsequently married. She said she hoped to move back to Taiwan, but her career in the NGO sector meant that better opportunities were available in London.

As can be seen from these examples, education was a common narrative thread among the UK cohort. Another such thread was expatriation – arriving in the UK as an expatriate or as the spouse of an expatriate, settling there, and (usually) establishing or taking over a small- to medium-sized enterprise. Yet another frequent pull factor was meeting and marrying, or forming a long-term relationship with, someone in the UK – not necessarily a British person, but someone predisposed to remain in the country. While London was not always seen as the most desirable place to live, and people often expressed a wish to move on someday, the amenities of a global city were often cited as a reason to stay in the UK. Even for people living and working outside of London, the city and its connections to the European market were a strong incentive to remain in the country.

London Chinatown: The Site and the Cohort

The London Background

London's appeal to skilled labour migrants is clear. As the UK's political and economic capital, London has long been considered a transnationally engaged location, with Sassen (2001) featuring it as one of the three archetypical "global cities" in her seminal work on the subject (New York and Tokyo being the other two). Having risen to prominence during the Middle Ages (Jenkins, 1988; Borer, 1977), London achieved a globally significant position during the time of the British Empire (Borer, 1977; Holmes, 1988), a period that saw the development of large communities of migrants in the area: economic migrants and refugee populations from within and outside the Empire, and business people seeking to profit from access to the British and/or Imperial market (Holmes, 1988; Berghahn, 1988). In the decades since the British Empire dissolved, London retained its position as a global city through the complexities of transnational finance: it has a lax taxation regime, and because of its geographic location, the trading hours of its stock exchange overlap those of New York and Tokyo. Also, between 1975 and 2020, its location in an English-speaking country within the European Union made London an obvious base for American companies seeking to invest and/or operate in Europe (see Courtney and Thompson, 1996, Introduction; Lewis, 1989; Thrift, 1994).

Despite the loss of business to European cities such as Frankfurt, Dublin, and Milan since the 2016 "Brexit" referendum on leaving the EU, the City of London remains a major hub for banking and advanced business services, as well as a locus of auxiliary services such as information technology. According to 2018 statistics quoted on www.cityoflondon.gov.uk, London contributes £41.8 billion per annum to the British economy and directly employs 513,000 people. The working population of London more generally is given as 26 per cent of non-European origin; by contrast, the figure for the UK overall is 10 per cent. London also retains a prominent reputation in art, design, theatre, and film, which provides further global connections besides enhancing the city's reputation. It also has a number of well-respected universities, including the members of the University of London Group, and good transportation connections with the universities of Oxford and Cambridge. This makes for a well-educated international talent pool from which global companies can draw, as well as a number of transnational communities of artists and scholars. London also continues to host refugees, asylum-seekers, and dissidents: the fact that Sun Yat-sen had spent part of his exile from China in London was a source of some pride for some in the local Taiwanese community, and at one point I was invited to a small memorial service held in front of the blue plaque commemorating his stay, which involved the placing of flags and bouquets of flowers and the singing of the Taiwanese National Anthem.

The UK generally, and London in particular, is also a magnet for foreign students. In 2013–14, the UK Council for International Student Affairs reported 310,190 non-EU students in higher education in the UK, rising to 435,495 when EU students from outside the UK were factored in (www.ukcisa.org.uk). While UKCISA did not provide a breakdown by city, London clearly attracts a significant percentage of these: University College London has the largest overseas student population, with 11,850, and the UKCISA's list of the top twenty universities by international student population includes five other London institutions (www.ukcisa .org.uk).

Relative to the UK as a whole, however, London is something of an outlier in terms of extent of, and tolerance for, migration. A 2009 report for the Migration Policy Institute indicated that the UK has only recently become a country of net immigration, with economic surges and the enlargement of the EU having boosted migration since 2004 (Somerville, Sriskandarajah, & Latorre 2009). Aside from a brief period following the Second World War and another during the New Labour government of the late 1990s, UK immigration policy has focused largely on restricting or limiting immigration, including in recent years (ibid.).

As a consequence, London is facing challenges to its status as a global city. The UK as a whole has seen a rise in nationalism and populist xenophobia since the 2008 financial crisis, the election of Conservative governments in 2010 and 2015, and the 2016 referendum in which the UK voted to leave the EU – although, significantly, 60 per cent of Londoners voted to remain (Koch, 2017; Devinney & Hartwell, 2020). An exodus of the financial institutions that have long underpinned London's globalization is currently under way (MacAskill & Davies, 2017). Changes to visa regulations have made it increasingly difficult for people outside the EU to acquire work visas compared to the 1990s and early 2000s (Warrell, 2015), a fact that has had a particular impact on expatriates, and also on students who are seeking to work in the UK after completing their higher education (Weale, 2015). Younger informants would mention in interviews the difficulty of obtaining, and keeping, a work visa, and saw it as an ethnic issue:

> As a Taiwanese it's not so easy to get a job, even the UK government only gives graduates under a two-year working visa to stay in the UK, but even so not so easy to get a job. So like the graduates from London University, they create their own business ... I am a Taiwanese and also I'm the President of [business networking association], if I do not pass the business to them I pass my business to the British people. (Michael, lawyer, 30s)

Other issues, such as the tightening market for affordable housing (Elledge, 2015), are also undoubtedly affecting London's attractiveness as a destination for skilled labour migrants. Yet the city remains a draw for the ambitious and the committed and remains more pro-immigration than much of the rest of the UK (Hill, 2015). Relative to other large cities around the world, it can still be considered a pre-eminent global city.

The Taiwanese in London

London is ethnically diverse by British standards; in the 2011 census, 55 per cent of the city's population of 8.2 million self-identified as non-white, compared to only 14 per cent of the total population of England and Wales (BBC, 2012; www.ons.gov.uk). London also had the highest percentage of residents born outside the UK, at 37 per cent, and of non-UK nationals, at 24 per cent (ibid.). While population statistics were unavailable for Taiwanese people living in London specifically, the British Office of National Statistics gives a 2014 figure of 37,000 people living in the UK who were born in Taiwan, compared to 196,000 born in mainland China (www.ons.gov.uk). The Taiwanese figure may in fact be slightly higher, for it does not take into account that Taiwanese people born before 1949 may have been registered as born in China.

London's Taiwanese community has no geographical focus apart from a loose affiliation with London's Chinatown and, in some cases, a social focus on the Taipei Representative Office, the main consular organization in the UK (as the UK does not formally recognize Taiwan as a separate country, Taiwan does not maintain an embassy or consulate), although how much of a focus these were appeared to depend on age, profession, and political affiliation. At the time of the initial London research, the KMT held a strong political majority and dominated Taiwanese politics, meaning that the Taipei Representative Office, and several Taiwanese business organizations, tended to present the KMT's political positions in the most favourable light. After the party lost much of its majority in the 2012 elections, these discourses were much less dominant. This influenced my study in that many of the people I had contacted through the Representative Office in 2009 and 2010 were either open or tacit KMT supporters, which influenced the views of Taiwanese ethnic identity to which I was initially exposed.

Members of the community would also meet on occasions such as New Year's parties or the annual Taiwanese Food Festival. Other loci of organization were provided, in the absence of physical ones, by online groups such as London Taiwan Town and the Taiwanese Student Association, which maintained Facebook groups. It should also be noted that Taiwanese living outside London maintained connections with the London community through organizations such as the ones named above, networking organizations such as the European Taiwanese Chamber of Commerce, and wider diaspora Chinese organizations for students, business people, and others.

The Taiwanese as Sojourners

Taiwanese migrants in London, even those who were or intended to become long-term or permanent residents, tended to speak of the city less as a permanent home than as a place to spend a few years building one's career before returning to Taiwan or moving elsewhere. An example is Maria (introduced earlier), an entrepreneur in her fifties who had in fact settled in London:

> Why [did] I choose London? ... I studied physiotherapy in Taiwan and then, during the university time, we had quite a few British lecturers, so of course they influenced me, and they said "Yeah, London is a good place to visit" and so I decided to come just to get more experience. And since I came to London I met my husband and I settled here ... He's also Taiwanese, but we met in London.

Maria, like other migrants, encouraged her children to maintain ties to Taiwan rather than regarding themselves as wholly connected to one country or the other.

> Oh yes, [my children] both love Taiwan very much. My daughter's in Taiwan at the moment. Yeah, even my son – they go to Taiwan at least once a year. Yeah, they love it. And they go by themselves – they have friends they have, they enjoy the food, everything in Taiwan, yes ... I think they feel they are both.

Kevin, another entrepreneur, had settled in the UK when his son was five, and his son had also studied abroad, in the United States, encouraged by his family. Kevin's son then greatly surprised his father by announcing that he did not want to inherit the family business and by asking to be released from the legacy. It is significant that his father had expected him to retain Taiwanese attitudes despite having been largely raised abroad; also significant is that the family still lived in an extended-family compound in traditional Taiwanese style, despite being in the UK and despite Kevin having released his son from the obligation to take over the company. Even among individuals who might be expected to integrate, it was expected that this would happen in only a limited way.

Given the previous discussion of the role played by international education in the London economy, it should be no surprise that a lot of recent migration to the UK has been student-driven. More interesting perhaps is that this has been encouraged on the Taiwanese side. The Taiwanese and British governments in recent years have actively encouraged study in the UK, as reflected in this excerpt from an interview with Mr Lee, then director of the Taipei Representative Office's Culture Section:

> The Government tried to encourage students to go overseas, and in the past many went to the United States, about 46,000 per year ... The UK is a new destination. [The number of students] reach[es] 20,000, but why do they come here? I think many [come here] because they spend only one year then they can get a Masters', [or else] because the family cannot reach, and so they cannot get a US visa, then they come here. [Interviewer: Is it easier to get a European one?] [*nodding*] Because the British Council, they work very hard, I think they open the gates.

This bears unpacking on several levels. First, the British government's stated policies are encouraging sojourner migration patterns, welcoming people from favoured locations for short stays but discouraging more settled migration or more wide-ranging recruitment. Second, Mr Lee's response suggests that some of the students, at least, may want to stay in the UK on a more long-term basis but find it difficult or impossible to do so. Third, the UK is a less desirable destination than the United States, in part because it is more difficult to get a longer-term visa (it is unclear whether his reference to family refers to the difficulty in bringing the family over or, more likely, the prohibitive cost of sending a student to the United States). Third and finally, it highlights that the Taiwanese government also encourages a Taiwanese presence abroad within the constraints of the host country, by supporting foreign student travel and openly sponsoring Taiwanese student organizations in the UK.

Student migration had a not insignificant impact on the migration of other Taiwanese, according to Mr Lee's account of his relations with the British Council: "So in the past, the last decade, the British Council, they went to Taiwan and come to our Ministry of Education and say 'We want to have ... we want to get into this market' ... So they get there and set up the office in less than fifteen years, yeah. So I think the students bring the [other] Taiwanese here."

Evidently, the dynamics of home and host country are involved in this focus on student migration. The British are less interested in attracting the students themselves as potential contributors to the British economy than in accessing what they see as a lucrative market. Whatever the case, they are encouraging the migration of talented Taiwanese to London through the connections the students forge during their stay. The student-and-expatriate-focused nature of the Taiwanese community encourages a sojourner rather than a settler approach to migration in London – something that is furthered by the British migration infrastructure.

Even so, the Taiwanese community would also (as Mr Lee's observations suggest) help people who wanted to settle, through ethnic ties: "When you find some difficulties, and no matter in your personal life or your business, you need someone to help you but sometimes, when you stay in a British company, you want to press your colleague to help you, or you can press your Taiwanese good friend to help you" (Michael).

The Taiwanese in London are thus constructed as sojourners seeking to develop self-help networks in order weather a sometimes hostile British social environment. However, their experiences also highlight the complicated interplay between home and host country politics, and this influences the strategies of individuals as they migrate between global cities.

The Taiwanese and the Chinese

One can expand on the complex relationship between the Taiwanese and British immigration systems by considering the equally complex relationship between the Taiwanese and London's wider Chinese community. Mr Lee spoke of how relations between Taiwan and the mainland were replicated in the London community: "Before that [i.e., in the twentieth century] I think some of the early immigrants, they come to here and stay and have their Chinatown, it's a long story ... But at that time China is part of Taiwan, we still had the dominant power, but later on China is getting stronger and stronger, then they shift [in status]."

As noted, London's Taiwanese community does not have a physical core but instead revolves around institutions. By contrast, London's Chinese community has an actual Chinatown. Other Chinese social spaces exist (for instance, a shopping centre in Croydon, and a multicultural suburb in southeast London), but these are less of a focus for community activities. However, the London Chinatown's connection to the Chinese community is complicated. On the one hand, it reflects migrants' efforts to reimagine Chinese culture in a European context, with the visual tropes of buildings and decorations recalling the kind of idealized Chinese past that, in terms of Taiwanese identity, has been something of a historical bone of contention between the mainland and Taiwan, as regards the KMT's claim that Taiwan is a repository of "pure Chineseness" (Murray & Hong, 1991). It is worth noting that at the time to which Mr Lee refers in the previous quote, Taiwan itself would have been a Japanese colony, and that he also refers to China as being "part of Taiwan" rather than the other way around, underscoring earlier observations about the KMT narrative regarding Taiwanese identity as "Free Chinese."

On the other hand, Chinatown has maintained connections with other groups. Shops in the area sell Chinese goods aimed at tourists (such as jade jewellery, lanterns, and figurines of "lucky cats," a symbol of Japanese origin that has spread across Asia), or non-Chinese Asian imports aimed at fashionable young Asians and Asianophiles seeking trendy goods from all over East Asia, such as Hello Kitty products and Japanese comics. London foodies of all ethnicities visit the Chinatown supermarkets looking for precious ingredients. The same was also true of the above-mentioned Croydon centre. The cosmopolitan nature of London is such that its Chinatown is both a nostalgic and colonial representation of a specific culture and a transnational, tacitly pan-Asian, space with connections to various groups, who make use of it in various different ways and engage in active discourse over what it represents.

Because there are relatively fewer Taiwanese in London, they have been compelled to build connections with other groups. One of my interviewees, Penelope, a woman in her seventies who was active in the Taiwanese Buddhist community, confirmed that the Taiwanese will use more generally Chinese-focused community organizations, such as the Chinatown Community Association, for support and to build connections. She herself worked in immigration services, providing chaplaincy for new immigrants of the Buddhist faith, not just from China but also from Malaysia, Vietnam, Sri Lanka, the Philippines, and India. To access social services and other benefits, people will build connections not only within their own community but also outside it.

For many, furthermore, London's main draw is its global cosmopolitanism:

> So why do we prefer to stay here ... When we stay in the UK, especially in London, there will be, like, multiple cultures; you can try so many different [things from] multiple cultures, like a museum or restaurant, and also you can be multicultural with friends from different countries. Yeah, it's different ... Probably, we cannot get it in Taiwan. I think that would be most of the reasons why the Taiwanese want to stay in the UK. (Michael)

Some aspects of Taiwanese life in London are common to all sojourner groups, albeit with a Taiwanese focus, such as language groups, traditional community celebrations, student groups, and the development of business and trade associations. At the European Taiwanese Chamber of Commerce conference that I attended, the seminars and business meetings were supplemented by distillery tours, city tours, karaoke contests, and a day-long golf tournament, all of which were seen as crucial to business networking, and none of which are unique to the Taiwanese. The Taiwanese do engage in internal networking through ethnic identity, but this itself leads to the development and maintenance of complex external connections with the host society and with other migrants.

The complex crossing of identities is evident in Maria's account of her involvement in setting up a networking association:

> That's why we started the Taiwanese Women's Association about ten years ago, yeah. We really try to tell everyone that we are Taiwan, and although the Taiwanese culture may be slightly influenced by the Chinese culture, it's quite different, yeah. That's why we started this. Although we already had a Taiwanese Association, and that we used to belong to the Taiwanese Association, but we thought yeah women will have different needs, yeah. So we still are members of Taiwanese Association, but we like to have one of our – women's own.

Here, Maria is asserting the distinctiveness of Taiwanese culture while acknowledging a connection to Chinese culture. She also combines gender with ethnic identity in her efforts to develop an organization to support Taiwanese women specifically. Networking also goes on *between* organizations: the head of a political organization aimed at unification with the mainland was also a key volunteer at a Taiwanese Buddhist temple, and again, both activities would have led to connections with other non-Taiwanese groups.

Maria's narrative of how she came to run her current business illustrates both the sojourner-focused nature of the Taiwanese in London and the way in which Taiwanese identity combines with other identities in the course of building a transnational community:

> So then of course I joined quite a few Taiwanese activities [and] communities. At that time there was a Taiwanese Travel Agent ... That travel agent [was who] we used to buy flight tickets, everything from that travel agent. Then that agent decided to stop trading ... Because I think [it] was run by a father and daughter, and then the father died, and then it became too much of a burden to the daughter, and she thought it was too much hard work, so she decided to stop it. Then we thought, yeah, [there are] not many Taiwanese in London but [they] still needed a travel agent, so my husband and myself, we decided to step in.

Given the combination of Taiwanese and cosmopolitan identities under discussion, it is worth noting that the travel agency focused largely on visits to London by prospective students and their families to attend interviews and evaluate universities; followed by tour groups from Taiwan; followed by visits to London by the relatives of Taiwanese living there. Maria also had built up significant connections with the Taiwanese business community and arranged travel and conferences for many local Taiwanese business associations. Furthermore, she used her connections both within and outside the community to identify consumer needs and develop her business. Her business may appear to be small and local, but it is driven by global markets and politics, connected to ethnic and other networks, and focused on the sojourner migrant experience.

In sum, the Taiwanese in London have constructed themselves as sojourners providing services to other sojourners. Even Taiwanese who have settled in London for the long haul see themselves as sojourners, whatever the reality might be. They also maintain ethnic networks that are combined with other identities to produce complex transnational and transethnic connections. It is worth, next, considering whether the same is true in a North American context.

Chapter Five

Toronto: The City of Settlers

The Taipei portion of the study came before the Toronto portion. Even so, I will consider Toronto next, for it is also a site to which the Taiwanese community has dispersed. Toronto is worth juxtaposing with London because both are "global cities," albeit with quite different ways of being global, and those differences have had significant repercussions for the two Taiwanese communities at hand.

A New Site: Background and Context

The Shift to North America

In 2013, after I returned from Taipei, an opportunity arose to conduct follow-up work in Toronto. During the London portion of the study, I had renewed contact with some family friends of Taiwanese descent in Canada, and when I was able to obtain funding for a short period of research, I decided to focus on the Taiwanese community in Toronto. I hoped to build a comparison between two cohorts of the same migrant group, one in Europe and one in North America, to consider whether, as many of my interviewees suggested, the migrant experience was categorically different in the two locations, and what this signified for the uses of ethnic identity.

Through one of the family friends, I obtained an introduction to a Toronto resident of Taiwanese origin who was well connected with the local community through her personal network and through her role in an international alumni organization for a Taiwanese university. As in London, I then resorted to the "snowball method" to find further interviewees, through the community leader mentioned above and other friends and family members and by contacting Toronto business networking organizations, although those organizations proved to be less forthcoming than the British ones in terms of providing contacts. Indeed, unlike in the UK, attempts to contact political and diplomatic members of the Taiwanese community in Toronto proved to be entirely fruitless. As in London, research in Toronto is ongoing; I continue to conduct interviews and participant-observation on a semi-annual basis during visits to the city.

The Toronto Cohort

In Toronto, formal interviews were conducted with thirteen individuals. Twelve of them were connected, directly or indirectly, with "Mrs Wang," a key member of the alumni organization of a Taiwanese university that was a focus of professional networking in Toronto (although not all of these connections were members of the alumni association in question). Two were representatives of other Taiwanese networking organizations, these ones focused on business.

Interviews were still the primary method for gathering information about people's lives and experiences. With some of the subjects, further data were obtained through email correspondence. Informal conversations were also held with these informants and one other, at networking events or community sites, or, since Mrs Wang's personal network intersected with the researcher's, during social visits, and some contacts have been ongoing. Further interviews have been conducted with members of Mrs Wang's network who no longer live in Toronto, having moved to the United States or British Columbia. Seven of the interviewees had STEM careers, mainly in engineering; an architect, two real estate agents, an entrepreneur, a housewife, and a graduate student formed the rest of the group.

"Toronto the Good": The City and the Community

The Toronto Background

The Toronto region was inhabited by Iroquois for centuries. The City of Toronto was officially founded, as the town of York, in 1793 by John Graves Simcoe, as the capital of what was then known as Upper Canada (Benn, 2004). The city would go on to become the capital of the Province of Ontario after Confederation in 1867 (ibid.). Today Toronto is the economic capital of Canada as well as the country's largest population centre. Along with Vancouver and Montreal, it is one of the country's most globally engaged cities.

Because the Harper government (2006–15) put an end to the long-form census, accurate Canadian population statistics for the time of the study are unavailable. The 2006 census indicates that in that year, the City of Toronto was home to 8 per cent of Canada's population, 20 per cent of all immigrants, and 30 per cent of all *recent* immigrants. Nearly half of Torontonians (1,237,720 out of 2.79 million) were born outside Canada, and 47 per cent (1,162,635) self-identified as a visible minority (www.toronto.ca). The 2016 census indicates a slight shift upwards on all counts, to a population of 2.96 million, with 51.2 per cent born outside Canada and the same percentage self-identifying as a visible minority (www.toronto.ca). The 2016 census also indicates that 14,705 people born in Taiwan were living in Toronto and its suburbs, and 63,770 in the country as a whole.

Socially, Torontonians tend to speak in terms of a divide between the urban core and the suburbs, or "Greater Toronto Area" (GTA). The urban core is reputed to be liberal and diverse, while the suburbs have a reputation for being less ethnically mixed (although the statistics on www.toronto.ca indicate near-identical numbers of people identifying as a visible minority in both) as well as more politically and socially conservative. However, the actual situation is much more complicated; for instance, the largest Taiwanese enclave in the Toronto area is, in fact, in the suburbs. Canadian identity tends to be heavily focused on the immigrant experience (Carrington & Bonnett, 1997), with a culture of tolerance for migrants being a strong point of identification (Kelley & Trebilcock 1998), even if the culture is not always tolerant in practice. The decision by the Conservative Party, in government since 2006, to run an openly anti-immigrant political campaign in 2015 is frequently alleged to have been behind their election loss that year (Dyck, 2016). Significantly, most Canadians tend not to hyphenate their identity, referring to themselves, for instance, as "Chinese," "German,"or "Jamaican," even if their original immigrant ancestor was one or two generations back. Notwithstanding this overt self-presentation as a tolerant society of migrants, there is a certain amount of tacit xenophobia, as well as both covert and overt discrimination, which has been consistently observed over the last few decades (Sharma, 1982; Carrington & Bonnett, 1997; Salaff, Grieve, & Ping, 2002; Barber, 2008; Preibisch & Hennebry, 2011). There are also possible signs that anti-immigrant feeling has been rising over the past decade (see Vomiero & Russell, 2019). This was supported by my informants, who, while they asserted that most Canadians were polite and courteous, also spoke of experiencing microagressions as well as both deliberate and unconscious discrimination based on their ethnicity.

Another paradox, noted by Funnell (2014, pp. 137–141), is that while Canada is a large node in the Chinese diaspora and a popular destination for many Chinese and ethnically Chinese people (see also Salaff, Grieve,

& Ping, 2002), Canada is seen by the Chinese as lacking in international prestige. Funnell notes that Chinese Canadian stars of *wuxia* (Chinese martial arts) films tend to downplay the Canadian side of their origins (p. 137). Canada's globalization, and its significance to the Chinese diaspora, tends to be tacit, so while it often features in literature on Chinese migrants, the implications and associations of the Canadian context are usually unexplored.

A Different Globalization?

Toronto is classified as an alpha global city by the Globalization and World Cities Research Network (www.lboro.ac.uk/gawc/world2012t.html). However, its claim to globalization rests on different grounds than London's. Toronto was originally a manufacturing city, but its industrial base has declined in recent decades in favour of service industries. The city today is home to a transnational financial industry with strong ties to the United States (Parnreiter, 2013, p. 27). Its history as a destination for migrants and refugees of all backgrounds (Carrington & Bonnett, 1997), combined with the ethos of the "Canadian Mosaic" (Kelley & Trebilcock, 1998; see also Berry, 2011, p. 2.4), which encourages arrivals to integrate into the host country even while maintaining ties to the home country (as contrasted with the American "melting pot" metaphor, which encourages migrants to assimilate and abandon their culture of origin; see Berry, 2011, p. 2.4), has resulted in a multicultural city with sharply defined ethnic communities and a number of active transnational connections.

Toronto consequently has a reputation less as a global business centre than as a multicultural space with a reputation for openness, tolerance, and transnational social connections (hence nicknames such as "Toronto the Good"), although this tolerant reputation was problematized by my informants (see below). My interviewee Lucille (50s, real estate agent) summed it up as follows: "I think I came to Toronto because it probably is much open and we have all different type of ethnic groups, and we are pretty tolerant in this society and it's very interesting, so they do respect this different cultures, so I found that it's very, just suit me, yeah, very nicely." Toronto's proximity to the United States also meant that some of my informants pursued transnational careers and social lives across the American border; for instance, some had family dispersed across the continent, or working in both Canada and the United States.

The Canadian political infrastructure is currently more hospitable toward migrants than the British one. For instance, Canada offers relatively easy access to visas for family members of skilled migrants, as well as work visas for graduates of Canadian post-secondary institutions (www.cic.ca). A Migration Policy Institute report describes Canada as having immigration embedded in its national identity. The policy focus in recent years has been on attracting younger skilled immigrants to replace an aging workforce (Challinor, 2011). Canada's immigration department maintains a relatively merit-based system, one that rewards education, fluency in French and/or English, and work experience over nationality (*The Economist*, 2015). Berry's studies of attitudes toward multiculturalism in Canada found that on the whole, both immigrant and non-immigrant populations view immigration and immigrants positively, though it is possible that attitudes have been hardening in more recent years (see Vomiero & Russell, 2019). On Toronto public transport, at the time of the study, one could see advertisements for immigration lawyers with testimonials from satisfied settlers, whereas in London, one was more likely to see ads for services helping people to obtain visas to the United States. Perhaps unsurprisingly, nationalist discourses in Canada have focused more on concerns about the loyalties of dual citizens (Black, 2015), or on encouraging selective immigration (Ibbitson, 2014), or on the image of the "ideal immigrant" (Barber, 2008), rather than on a fully closed border, as the equivalent discourses do in the UK. In contrast to the UK Conservative government's rhetoric about immigrants taking the jobs of British workers (thus largely conflating temporary labour migration with more long-term immigration), Canadian Conservatives have focused quite openly on attracting economic immigrants, differentiating between short-term labour migration and long-term settlement (Ibbitson, 2014; *The Economist*, 2015). While the process of sponsoring family members for settlement has become more difficult in recent years (see www.immigrating-to-canada.com), it is not openly "hostile" like the British equivalent (Redclift and Rajina, 2019). Chiang's (2011) study of return migration from Canada to Taiwan, while unable to provide detailed statistics, found that most informants had returned for family or employment reasons rather than for reasons of nationalism or xenophobia. So while nationalism does exist in Canada, its relationship to immigrant communities is rather more complicated and less openly hostile in social and policy terms.

At the Pacific Mall: The Taiwanese in Toronto

The Taiwanese Community

The Taiwanese community in Toronto is more geographically cohesive than the one in London. There is a heavy concentration of Taiwanese and ethnic Taiwanese in Markham and other eastern outer suburbs of

Toronto, with a geographic and social clustering around a shopping centre called the Metro Mall, which houses primarily Taiwanese businesses and organizations. There is also the Pacific Mall, also in Markham, whose shops ostensibly cater to all Asian groups (albeit with a certain dominance by the Cantonese). Among some informants, however, the use of ethnically focused business organizations, and Taiwanese student and alumni associations, remained crucial.

Like the UK, Canada does not officially recognize Taiwan as an independent country, instead supporting mainland China's claim to political sovereignty in the region. However, Taiwan retains an independent diplomatic presence in Canada through the Taipei Economic and Cultural Office in Ottawa (analogous to the Taipei Representative Office in London), as well as an extensive economic presence through trade and expatriation. Also, Taiwan is unofficially recognized as an independent country in Canadian popular culture.

The Academic Migrant: Mrs Wang

Mrs Wang is from a family that migrated to Taiwan in 1949 but that does not identify as *waishengren*, for they had migrated for business reasons rather than out of ideology. After attending the National Taiwan University, she came to Toronto to do a postgraduate degree in chemistry; she had intended to study in the United States, but at that time that country maintained quotas on the number of student visas per Taiwanese family. Canada was her second choice. Her husband, also Taiwanese, came to Toronto to study for much the same reason. She worked in academia for a time before shifting to the IT sector. Now retired, she is very active in the NTU alumni network. She lives in an ethnically mixed but predominantly white mid-urban area in eastern Toronto. Her two grown children both live in Canada.

The Juvenile Settler: Kay

Kay came to Toronto as a teenager, along with his younger brother, when their parents migrated in the early 1990s; his older brother followed a year later after finishing his national service. His *benshengren* parents were self-directed skilled labour migrants, working in STEM-related professions; they have since moved back to Taiwan. Kay himself studied engineering, having grown up in a STEM family, and now works as a software developer. He lives in a suburb of Toronto that is predominantly mixed white and Chinese. He is not a member of any ethnic association and doesn't actively follow activities related to the Taiwanese community;

although he still celebrates Taiwanese festivals, eats Taiwanese foods, and enjoys Japanese-exported cultural activities such as karate and manga, there is a social expectation in Canada that people – recent immigrants particularly – will want to celebrate the festivals of their ethnic groups, and Japanese popular culture is pervasive. He is married to an ethnically Hong Kongese woman; they have two children and a very ethnically mixed circle of friends.

While settlement after going abroad for higher education, as in London, was a common theme among the older members of the cohort (the ones who immigrated before the end of the dictatorship in the 1980s), younger members had generally migrated along with their families (or they had been born after their families settled in Canada). This may, however, be an artefact of my sample, for it has been well documented elsewhere that in Asia, Canada is currently a popular destination for foreign students (Teo, 2011). There was also a noted focus on the STEM professions, at least as an initial career choice, among both men and women (though, significantly, the exceptions to this were all female). There were fewer expatriates and entrepreneurs in the Canadian sample; most of these people worked for existing firms or in the public sector after graduation, which, as detailed below, is most likely a consequence of the different employment market in Canada. Another significant point is that Canada was not the first-choice destination for any of the migrants; all had intended to move to the United States and had come to Canada (or remained there) because, for various reasons, they had been unable to enter the United States.

Taiwanese as Settlers

In contrast to the sojourner mentality among Taiwanese in London, Taiwanese migrants in Toronto saw themselves as permanent settlers, even if this was not always the case, and even if they maintained transnational connections. Unlike in the London cohort, there was an expectation that their children would assimilate rather than maintain transnational connections. One interviewee in his thirties, Kay, said that if his children (currently preschoolers) showed an interest in Taiwan later on, he would encourage them to explore their identity in that direction, but that if they didn't, he wouldn't push them to do so. Similarly, Johnny, a retired engineer, told me that "my sons, they got a different culture, they got different thinking, because they all born here." There was a physical centre to Toronto's Taiwanese community, reflecting the usual Toronto practice of informally dividing the city into ethnic "villages" or enclaves, and this had resulted in a stronger "home" location for the Taiwanese, even if not all members of the community lived in the area.

London was an attractive destination in and of itself for Taiwanese migrants, whereas Toronto was generally not the first-choice destination for the people with whom I spoke. Of the migrants who were not brought to Canada by their parents, most had intended to go to the United States and had either come to Canada with the intention of moving south or had chosen Canada as their second-best option on learning that they could not migrate to the United States:

> Why Canada? That's a good question. OK. My mum's friend's son got an application form, so he doesn't need that one, he gives it to me. So I applied. I got a scholarship, so I came. At the same time I had a University of Tennessee entrance [offer], but they gave me an offer of winter season, so my mum says ah, you just go to Toronto, if you don't like it you can go Tennessee in wintertime. And I did. It's OK! (Laura, real estate agent, 60s)

Even when Toronto was the first choice, they may not have been initially aware of it as a possibility:

> Well, actually, the one who wanted to emigrate in the beginning was my daughter ... [She] started to feel the pressure of, of joint examinations [at] high school and that was the year she starting telling me that she would like to come to the USA, or other countries. Then I say, "Okay, let's find a company for immigration and ask," and the first time we went to listen to the speech [*laughs*], the company, I swiped my credit card [*laughs*] because they are so good in talking about those education system ... of course, Australia or New Zealand were possible countries, but I think North America anyway is mainstream, alright [*laughs*]. (Julie, retired, 50s)

It is worth noting that Laura had migrated in the 1970s, whereas Julie was a relatively recent arrival, having migrated in the 2000s with the aim of furthering her daughter's education, in line with what another interviewee called the "Tiger Mom" pattern, of mothers who put their children's educational needs at the forefront of family priorities. While interviewees generally cast their migration to Toronto as a fortuitous event (one informant had subsequently moved to the United States, before moving back later on the grounds that society was less tolerant there), Toronto stood in contrast to London, which was generally seen as an attractive destination in and of itself for migrants. Ironically, then, the city with the "settler" discourse for migrants tended not to be viewed as an ideal destination, while the one with the less friendly "sojourner" discourse was.

The Taiwanese in the Multicultural Context

The Taiwanese in Toronto defined themselves as a distinct community relative to others in the city, using much the same symbols of identity as the Londoners did, such as the use of Mandarin as a spoken and written language, as well as particular types of food. One young interviewee demonstrated this clearly when talking about how she differentiated her "Taiwanese" (meaning Taiwanese Canadian) friends from others:

> I still do hang out more with my Taiwanese friends than my university friends. A lot of them came at the same age as I do, so sometimes we even speak just English but still we, you know, more than, um, yeah ... I'm not quite sure, just because I know them since from I was so young in high school or because it's the Taiwanese thing. I think there are, in terms of culture, of what we enjoy in our off time, there's a certain difference. So let's say people here, when they have a conversation they always, almost always, start with music or something like that. But I'm not very familiar with popular music, and I don't really go hang out in a bar or go to clubs, so they'll say, even just to, just to break the ice with some people here, it's not as easy. But I guess people in Taiwan, the first thing they talk about maybe is food or ... It's either food or try to ... or try to connect like they're, first of all they're trying to figure out if you guys know somebody in common, let's say, and then sometimes you do like from the farthest end and it's like some ridiculous little thing and somehow you know "oh I know that person," "oh I know that person too," that kind of thing. (Lucy, architect, 20s)

This seemed to be the case even among people who considered themselves fairly well assimilated into Canadian society, such as Kay, who, like Lucy, had migrated to Canada as a child with his parents and sibling. He indicated the markers of Taiwanese identity even as he asserted that he identified less strongly than others as Taiwanese:

> I think relative to other Taiwanese, I think I've been assimilated more than other people, I am even starting to forget how to speak Mandarin, but I still consider myself Taiwanese more than Canadian in terms of who I identify myself as. Language-wise, it's just because practically I don't have anyone to speak Taiwanese to or Mandarin to because all my families are actually not here with me and I hang out mostly with [my wife] and her family and they're Cantonese in origin, so I speak in English with them but I still identify myself more of a Taiwanese than Canadian, yeah. (Kay)

The Toronto Night Market, like London's Taiwanese Food Festival, was also a hub of community identity, and it is worth noting that other Canadian cities with significant Taiwanese populations, such as Vancouver, also have regular night markets. Local ethnic business and community organizations, and locations such as the Pacific Mall and the Metro Mall, were similar foci of activity. Toronto thus maintained a Taiwanese community employing symbols of identity, tropes, narratives, and discourses analogous to those in London.

Like the Londoners, the Toronto Taiwanese maintained connections with the rest of the city's Chinese community. But in Toronto, this relationship could be somewhat more tense than in London: two older interviewees, one of whom was very active in the Taiwanese community, complained about the Toronto Chinese community being dominated by the Cantonese, who, since they were the main presence in most of Toronto's Chinatowns, also had the greatest visibility (significantly, when non-Taiwanese Canadians found out that I was taking Chinese lessons, they often assumed that I was learning Cantonese, whereas Londoners assumed I was learning Mandarin). Lucy, too, was slightly disparaging towards the Chinese Canadian identity: "The Chinese Canadians are interesting. They ... sort of, they feel like they are Chinese but they're not Chinese and they don't, some of them don't know very much about Chinese."

The Toronto International Film Festival posters advertising its Chinese film festival, which were visible when I conducted a research visit to the Pacific Mall, were diplomatically subtitled "Mainland/Hong Kong/Taiwan," acknowledging three main strands of affiliation within the Toronto Chinese community as well as the three main hubs of Chinese filmmaking. There was a community support organization in Toronto that was specifically geared to the social welfare of Taiwanese migrants, whereas in London, the main Taiwanese-specific organizations tended to be aimed at businesspeople or students specifically; social welfare organizations were framed either as Chinese or in terms of other identities (e.g., Buddhist groups, or the Taiwanese Women's Association). As in London, connections were built with the wider Chinese community; however, these were often couched in local struggles for dominance and reflected a greater norm toward settling in the city.

Perhaps significantly for the image of Taiwanese as settlers, the Toronto cohort tended to complain more openly than the London cohort about experiencing ethnic discrimination. London's Taiwanese tended to refer to this allusively or in the context of political institutions making certain activities, such as acquiring a visa or setting up a small business, difficult for migrants, rather than in the context of interpersonal interactions:

"You look up a journal, most publications always have a surname, or an English name like Moore [laughs], and now [as second author] there is Chinese name, and now [as third author] there's Indian name, you know right away that English name is the professor [laughs], the other two are just the slaves that are working as post-doctorate or graduate student" (Mrs Wang, retired chemist and IT professional)

Kay, too, saw discrimination as a possibility, though in relation to Canada *beyond* Toronto:

> I do think Toronto is a mosaic, I mean it's very, it's almost, I feel that it's actually become part of Toronto's identity because I have heard, you know, friends ... who tried to find houses outside of Toronto and you don't have to go very far outside, and the further out, and he was telling me that the further out that he went, the more discrimination he felt there.

In part, this refers to the traditional divide in Toronto between the (reputedly) diverse, cosmopolitan inner core and the (reputedly) more socially conservative and ethnically homogenous suburbs; however, it also shows that Kay and his friends and family were connected with Toronto discourses of tolerance, integration, and cosmopolitanism. Arguably, this suggests that the Toronto cohort was more strongly invested in the local community than the London cohort was in theirs, in that discrimination was portrayed as an obstacle to be overcome or confronted rather than something to be worked around, or as something that existed for the most part outside of the city.

The Toronto Taiwanese also maintained pan-Asian connections, like their London counterparts. Steve (graduate student, 30), a representative of a local Taiwanese business networking organization, when asked about transnational business, spoke straight away about collaboration with partners from Korea and mainland China, before launching into an anecdote about setting up a business connection between two ethnic Taiwanese entrepreneurs, one in Toronto and the other in Vancouver. When Kay was discussing his self-identification as Taiwanese, he ultimately wound up broadening it to a kind of generalized Asian identity rather than a specifically Taiwanese one:

> But personally I do have more trouble identifying myself with non-Asian people, you know, communication-wise it's fine, but it's when maybe there's a cultural difference, and also how people behave or expect to behave in social situations, and at least for me personally I find it more difficult to fit into those crowds. Yeah, so that kind of drives me personally to kind of hang out with more Taiwanese or Chinese or Korean or Japanese people.

The Pacific Mall is openly pan-Asian. The Toronto Night Market, while Taiwanese-led, is also tacitly pan-Asian. Vendors there served predominantly pan-Asian food, along with other non-Asian ethnic foods such as Dutch *frites*, and various iterations of "fusion" cuisine, such as "Taiwanese burgers," North American versions of sushi, and *mochi* (a dessert popular in Japan and Taiwan, involving glutinous rice flour balls filled with sweet bean paste) with non-traditional fillings such as chocolate. The main non-food-related exhibitors at the London Taiwan Food Festivals were religious groups and a recruitment consultancy that specialized in recruiting young managers wanting to work in Asia; the ones in Toronto were local businesses and organizations with no Taiwanese connections, such as a car dealership, a toy vendor, a local bank, an insurance company, and representatives of a local public sector union (that was in the middle of a prominent labour dispute at the time of fieldwork). This would seem to link Taiwanese identity to wider Asian and transnational identities, besides showing that Taiwanese symbols could be hybridized with Canadian ones without losing wider transnational connections.

The Taiwanese in Toronto also used other identities than their ethnic ones to develop transnational connections. Johnny had experienced a transnational career, but one that had involved cultivating contacts in the United States as well as maintaining contacts in Taiwan, and one in which his professional career in engineering was the main driver. Lucille, a real estate agent, asserted that she seldom used her Taiwanese connections to get business, instead relying on advertising and other professional modes of advancement. Significantly, she also explained her personal reluctance to join Taiwanese business associations: "They are most probably tied to political [parties, i.e., the KMT]." In this way, she indicated a link between political affiliation and ethnic identity. The Taiwanese in Toronto used their transnational identity to develop multiple cultural connections, and combined their professional, ethnic, and other identities to do so, but along different lines than the British group.

The Taiwanese cohort in Toronto showed some similarities with its British counterpart, including shared symbols of ethnic identity, similar discourses relating to these, a tendency to combine ethnicity with other identities, and strong transnational connections. But they also tended to frame their experiences within a "settler" rather than a "sojourner" paradigm, assuming long-term or permanent migration whether or not this turned out to be the case. This was precisely the opposite to the dominant discourse in London. Comparing the two cities therefore shows that just as global cities can be global in different ways, so can a network society have different expectations and norms in different nodes.

Chapter Six

Taipei: The City of Origin

For methodological reasons, studies that consider a migrant labour network at its point of origin, in addition to its nodes abroad, are relatively rare, yet crucial insights can be gained from doing so. This chapter, therefore, will look at Taipei, how it came to be included in the study, how it stands up as a "global city" compared to the others, what the Taiwanese portion of the research contributed, and how Taiwanese culture affects and informs the construction of Taiwanese identity in the diaspora.

Room for Study: Taipei Background

How the Project Went to the Island

After the initial London-based section of the study was completed, the opportunity arose for me to extend the work through a short-term visiting fellowship at National Cheng Chi University (NCCU) in Taipei in the spring of 2013. While there, I was able to conduct a small number of interviews and, more importantly, gain insight into the Taiwanese experience by actually living and conducting academic work in Taiwan itself. As the ethnographic experience depends in large part on the researcher's ability to learn and analyse the tacit and unconscious social and cultural background to the subjects' activities (Sanday, 1979), this sojourn proved vital in terms of understanding the transnational Taiwanese experience. Some contact was also made with a member of the London cohort who conducted business in Taipei, and also with faculty of Sun Yat-sen University in Kaohsiung, where I gave a guest lecture. Interviewees were found through the NCCU MBA program and its alumni association. My three research assistants and several colleagues also contributed insights for this portion of the study.

The Taipei Cohort

In Taipei, formal interviews were conducted with only eight interviewees, as time for such activities was limited. This was supplemented with ongoing conversations with my three research assistants, who served as primary informants, and through many informal conversations with NCCU faculty members and graduate students, as well as the faculty and graduate students at Sun Yat-sen University. I also, as noted earlier, held a meeting with a London informant who had come to Taipei as part of his business activities; he is listed under the Taipei cohort in Appendix 1. All interviewees self-identified as "Taiwanese," with some also identifying as Chinese and one as Hakka.

Rising Tiger: Taipei as a "Global City"

A New Globalization?

Taipei developed into an urban conurbation in the eighteenth and nineteeenth centuries, although the area had been inhabited by the aboriginal Taiwanese prior to the period of European and Chinese colonization in the seventeenth century (www.english.gov.taipei). It was declared a prefecture city in 1875, when Taiwan was a Chinese province (ibid). The influence of the Japanese occupation can be seen in the city's architecture and in various social traits such as a fondness for certain foods of Japanese origin as well as (at least according to my informants) a stronger tendency toward politeness and submissiveness than one finds among mainland Chinese. After 1948, the KMT established Taipei as the "wartime capital" of the Republic of China (www.english.gov.taipei; see also Murray & Hong, 1991; Bedford & Hwang, 2006, ch. 1). Since 1949, Taipei has been the official capital of the ROC, the seat of government, and the island's economic hub; headquartered there are the various national media companies, 80 per cent of the country's leading manufacturers, and twenty universities (almost 40 per cent of the national total) (Wang, 2003, p. 317).

Wang (2003) examines Taipei's claim to global status, applying a variety of metrics and generally finding that it scores fairly high on most; he concludes that it is arguably the most global city in Taiwan, possibly even in Southeast Asia (Wang, 2003), on the basis of its role as a regional financial and industrial hub and as the country's political and economic capital (Wang, 2003, pp. 314–322). Socially, Taipei maintains a globalizing outlook comparable to that of London and Toronto: international consumer goods and designer brands are visible there, as are restaurant and retail chains such as Starbucks, Pizza Hut, The Gap, and Swarovski; all of these are especially popular among the young. I also noted a number of European restaurants, principally Italian (one informant asserted that this was due to Italian men marrying Taiwanese women and immigrating to Taiwan; I was unable to verify the accuracy of this statement). Tourism by non-Taiwanese is common; for instance, on a visit to the National Chiang Kai-shek Memorial Hall, one of my research assistants explained that most visitors were mainlanders who visited the memorial "out of curiosity." One of the videos that the Taipei Metro played in rotation in order to promote its services and explain safety procedures featured a white backpacker, "Sara" (implied to be European or North American), who has come to Taipei to visit her friend "Bill" and is impressed by the good manners of the Taiwanese. Besides reflecting a common Taiwanese point of identity (that they are more courteous and polite than mainland Chinese), it is considered normal to have North American or European friends and to entertain non-Taiwanese visitors.

Though it has attained global status, Taipei is not really a city of migration like Toronto or London. In 2014 its immigrant population was 60,749, with a further 86,604 in New Taipei City. The immigrant population for the entire country was only 621,757 (www.immigration.gov.tw). By comparison, Taipei's total population is 2,702,315 (www.english.gov.taipei). Most of the academic work on migration to Taiwan has focused on non-elite migration from less developed countries rather than on skilled labour immigrants (e.g., Hoang, 2016; Asato, 2017). Elite labour migrants and expatriates were not unknown to me, however: one interviewee for this project was a Swiss expatriate settled in Taipei; another was a Canadian who had spent an expatriate term in Taiwan; and two more, who were Taiwanese, currently or had recently worked with European and North American expatriates as local employees of the global automotive corporation MNC. There was also my above-mentioned contact who operated in London and Taipei. Furthermore, most of the people I met at NCCU and Sun Yat-sen University had studied or lived abroad, mainly in Japan, Europe, Scandinavia, and/or the United States, and at least one was in an ongoing long-distance relationship with a European. However, Taipei seemed to lack the active multicultural and globalizing ethos found in the other two cities. It is worth noting that "Sara" in the Taipei Metro video was quite plainly a visitor, not an immigrant. But it is also worth noting that the diaspora and the trend toward international study and visiting did seem to lead to connections around the world; migration is, after all, not the only form of transnationalism, and if the definition of "international manager" can include telecommuters and short-term expatriates as well as traditional expatriates (see Forster, 2000;

Crowley-Henry, O'Connor, & Al Ariss, 2018), then the same can hold true for transnational individuals more generally.

Taipei has retained transnational connections through its history of colonization by China, the Netherlands, Japan, and the *waishengren*, as well as through the current population of unskilled migrants from elsewhere in Southeast Asia seeking work opportunities in its cities (Hoang, 2016; Asato, 2017). One expatriate blog, www.taipeiexpat.com, describes Taiwan, facetiously, to outsiders as follows: "Taiwan is what you would get if Japan and China had a baby, and Japan raised it." Taiwan, like London, has a strong international education industry, with the government encouraging foreign students interested in learning Mandarin to come to Taiwan rather than the mainland (http://english.moe.gov.tw). Significantly, while the official number of foreign students pursuing university degrees in Taiwan in 2011 was 8,377, the number of foreign students overall, including in short-term programs, was given in the same year as 48,000, suggesting that a large proportion of these students were on short-duration language courses (www.taipeitimes.com). The global rise of Taiwanese companies, especially in the electronics sector, has encouraged global business connections; so has the increased outsourcing of North American technical and software industries to Asia. Finally, the diaspora seems to have influenced the mother country – Canadian companies such as the clothing retailer Roots and the cafe chain The Second Cup, neither of which are common global brands, maintain a visible presence in Taipei. (According to my younger research assistants, Roots has become popular enough to inspire counterfeiting.) Taipei, at the time of the study, was a city with both visible and invisible global ties; but it was also more a city of expatriates than a city of settlers or of sojourners, and/or a city with a strong awareness of its country's diasporic and colonial connections.

The Digital Globalizer: Mrs Wong

Mrs Wong was educated at NCCU in Taipei and started working in computing in the 1980s, during which time she had two children. In the late 1990s she started her own computer animation company, specializing in work for the cultural and heritage industries, such as animating educational programs and creating interactive displays for museums. She was motivated to start her own company by her search for her own identity, in that she had grown up in Taiwan being told to think of herself as "Chinese," only later to discover her independent identity as Taiwanese. Also, she wanted to give her children a sense of identity that she herself had lacked as a child. While she has never been an expatriate, Mrs Wong takes commissions from all over the world (albeit mostly from Asia) and often travels to places like Africa and South America in the course of her work.

Because of the smaller sample size in the Taipei case, I will include only one example here. Most people in the Taipei sample had not gone abroad for work, but most had travelled extensively, and a few had studied abroad, in Scandinavia or the UK. Most, furthermore, were in work that involved an international clientele or international managers, engaging with these people using information technology, sometimes supported by short visits. Most of the sample were in the IT sector in one way or another, though, as Mrs Wong's case indicates, the forms this takes could be surprisingly diverse. Finally, it is worth mentioning that the sample included one European skilled labourer who had migrated to Taipei and settled, to provide the reverse perspective.

Being Global Locally: Taipei

The Taiwanese at Home

While the Taipei cohort were settled in Taipei and not actively engaged in migration, all of the people I interviewed or worked with in this cohort had global reach and connections, even the ones who had not themselves travelled abroad. All worked for companies or organizations with international or overseas aspects to their work (one had a side business as a tour operator), had friends and/or family overseas, and made active use of the Internet to keep globally connected and in touch with both the Taiwanese diaspora and the wider world. At least one, who worked in IT, saw this as connected with Taiwanese identity:

> In Taiwan, [the] game industry, it's considered an advanced IT industry and for people in Hong Kong and for the people in China they like this advanced IT, whatever, and they all think that this advanced technology from Taiwan are good ... Colleagues that left the original company, they left for China, they can actually find really good jobs because they worked in [a] Taiwanese IT company. (Marina, online community manager for gaming company, 40s; interviewed in Mandarin)

Some of the symbols of identity used by the Canadian and British cohorts were clearly echoed in Taipei. The diaspora's focus on food as a symbol of Taiwan was, if anything, even more prominent in Taipei. Food was an ongoing conversation in Taipei, with discussions of new restaurants, or which venues were the best for particular dishes, being roughly equivalent to talking about the weather in Britain, and restaurants of various sizes and origins were ubiquitous throughout the city. Furthermore, in contrast to the focus on fusion cuisine evident in Toronto, international cuisine in general had tremendous social cachet, with European restaurants seen as exotic and sophisticated. Taiwanese identity in Taipei is linked to food but also to a culinary cosmopolitanism that ties into wider discourses of diasporic and transnational identity.

Discourses of Identity in Taipei

The pan-Asian discourse was a focus of identity for the diaspora cohorts but did not tend to manifest itself in Taipei. The Japanese connection was sometimes brought up, but this was generally in the context of its role as a former colonial power (or in terms of explaining the popularity of certain foods and restaurants). Vietnamese, Korean, and other ethnic groups, meanwhile, tended to figure mostly in discussions of exotic cuisine. The ambivalent relationship to the mainland was somewhat more visible:

> Starting from the seventeenth century, Taiwan was actually a quite cosmopolitan place because at that time you have Portuguese, we have Spanish and then Dutch. But then being a Chinese when I was young, well a Taiwanese, I was taught that I'm a Chinese and then, but then when I started to work ... it most all of a sudden changed, and people telling that we are Taiwanese. And by doing this [i.e., a project for a mainland Chinese heritage organization] for me is really to search for my own identity. (Ms Wong, 40s, owner of small IT consultancy; interviewed in Mandarin)

Ms Wong's remarks allude to the fact that in her lifetime, the discourse of Taiwan as "Chinese" had given way to the idea of Taiwan having a distinct identity, a development that fostered complicated and ambivalent attitudes in people born in the 1960s and 1970s. Similarly, Damon, a lawyer who specialized in international work, explained that the main business of international law in Taipei was bound up with the complexities of Taiwan's identity as both Chinese and not-Chinese: "We clearly have the focus of doing inbound work, inbound meaning from outside China/ Taiwan, Taiwan/China, into Taiwan and Hong Kong and China."

Marina reported that the gaming company she worked for had started out doing online mah-jong, a Chinese game whose popularity has spread outside the diaspora but that remains a Chinese cultural artefact by association. Many people in Taipei took pride in embracing aspects and symbols of Chinese identity that had been rejected or had fallen into disuse on the mainland, such as Taoist temple parades and the Imperial Chinese past (Mrs Wong's company was at that point working on interactive material for the National Palace Museum, a Taiwanese repository of treasures from Imperial China – whose displays, not insignificantly, tended to largely gloss over the separation of Taiwan from the mainland). The complex discourses of Taiwanese identity – they were both a separate group and part of a wider Chinese diaspora – thus took on new aspects in Taipei.

One aspect of Taiwanese identity that I encountered in Taipei, but that did not seem to feature much in the diaspora, was the idea that Taiwanese people are "hard-working": "Okay, yeah, okay, the Taiwanese people is hard to work [i.e. hard-working], yeah, also intelligent, but normally most of the time they don't want to follow the rules, they will be challenge the rule, say yeah, it's a shortcut to the target" (Frankie, local manager for European automobile company, 30s).

When similar discourses came up in the diaspora, it was not in the context of being Taiwanese, but in that of being an immigrant (or Asian more generally). Yet there did seem to be a tacit connection here with the diasporic tendency to blend professional identity with ethnic identity: various Taipei interviewees linked their ethnic identity to their work as IT professionals or transnational managers. Although the discourse may have had a different focus, then, the complexities in Taipei and in London and Toronto were similar with regard to the links between ethnic and professional identities.

There was also the idea, periodically expressed, that the Taiwanese were more straightforward or "more honest" than mainlanders or Hong Kong Chinese:

> Maybe they are the same in terms of their honesty, but I think they are more tricky and they think they are more superior than Taiwanese people, yeah, because especially for Hong Kong under, before it returns to mainland China and [pause] I think in some way they think that they are better than us, in some ways, because of their education, or in terms of the government, I think that they think they are better than us, yeah. (Jaime, entrepreneur, 30s)

This arguably linked in with the idea of Taiwanese being more hard-working, and possibly with the "pure Chinese" discourse, which set up Taiwanese people as morally superior to mainlanders.

One particularly interesting insight came from Damon, a Swiss-born European who had settled in Taipei after arriving as a young man to study Mandarin, and who had pursued a professional career in the legal industry: "In Taiwan it's, by definition it would be, very international, because everybody studies abroad, everybody goes abroad for vacation twice a year, so from the actual standard through ... they are thinking internationally or even acting internationally, but when it comes to news and all that and sort of the mindsets, it became very, very local, probably more local than elsewhere. I don't know why." Later in the same interview he added: "I think that the added value of [being] a foreigner in Taiwan is in being a foreigner ... And you're sort of making that gap, you're bridging that gap." From Damon's etic perspective, then, we get another view of Taipei as localized but outward-looking, as being not very transnational in and of itself but seeking to build and maintain transnational links, through the diaspora and through connections with foreigners.

Sarah, finally, described Taiwan's relationship with the global as complex and ambivalent. The global offered new opportunities to learn, but it was also a threat. There were echoes here of Canada's pro-immigrant discourses but also the UK's xenophobic ones: "They [Taiwanese corporations] will like to invite many foreigners, foreigner talent to join them, but always they make so many boundary for kind of people, so like a traditional Taiwanese company always can invite too many talents or talents won't stay there longer" (Sarah, IT manager, 40s). Taiwanese identity in Taiwan thus has globalizing aspects but also displays ambivalence toward immigration and emigration. It also reflects a complex discourse over the relationship between Chinese and Taiwanese identities.

Taiwanese identity as expressed in Taipei was ostensibly more straightforward than in the diaspora (one interviewee, when asked, "What makes you Taiwanese?" responded, "I was born here"). Its symbols and discourses were similar to those in the diaspora, and it also – again, as in the diaspora – evidenced clear transnational connections and a combining of identities (e.g., Chinese with Taiwanese, professional identities with ethnic with ethnic ones) to build transnational connections. Taiwanese identity in Taipei was global in different ways than in London and Toronto, and it involved different discourses of identity. This examination of Taipei indicates how discourses of identity at different nodes of a transnational network are developed and transformed as people flow from one to the other.

Chapter Seven

Cutting Bamboo: Migrants and Transnational Ethnic Networks

When Chinese migrants appear in the literature on transnationalism, it is usually in the context of writing about the Chinese diaspora and "bamboo networks." That term is the informal, somewhat romantic name for global networks of ethnically Chinese people, which are usually based on family though they are not necessarily limited to immediate family (Nonini & Ong, 1997; *The Economist*, 1995; Tanzer, 1994; Dahles, 2005; Tien & Luan, 2015). A related topic is *guanxi* networking, that is, networks of social obligation based on the exchange of favours. Both these networks, however, seem to be considerably more problematic on the ground than in the literature; they are hard to differentiate, linked with a variety of other networks, and heavily fraught with social and interpersonal tensions. All of this, again, links back to the concept outlined in chapter 2, of global networks being constructed in a space of flows between geographic nodal points, through networks based on ethnicity, religion, national identity, and other such traits, whose results are not always straightforward or positive.

In this chapter, therefore, I will consider how the Taiwanese skilled labour migrants used their ethnic identities to construct transnational networks focused on Chineseness and/or Taiwaneseness, and the extent to which they do, in fact, fit this paradigm of constructing "bamboo networks" and using *guanxi* connections across borders. I will, after that, propose that Chinese identity be viewed as a global rather than a national identity, albeit one whose meanings and interpretations are contested. The use of online networks and social networking sites, while arguably relevant here, will be considered in the next chapter.

All Networks Are Not the Same: Diaspora, Chinese Identity, and the "Bamboo Network"

Ethnic diasporas, and their role in transnational networking, have been a key focus of studies of migration from the earlier years of the discipline.

This focus can be tacit, as in studies of migrants in specific ethnic diasporas and the role their diaspora plays in their lives (e.g., Shaw, 1988, Banks, 1992, Björklund, 2003), or more generally, diasporas can be explored as a universal social phenomenon (e.g., Cohen, 1997). More recent literature has, however, been critical of the diaspora concept. Ho (2011a), for instance, argues that the term is overused and loosely employed and is too often hijacked by national governments seeking to develop a transnational presence by maintaining connections with their overseas migrants. Ndhlovu (2016), similarly, contends that the concept is overly simplistic and that attempts to develop more complex lenses have generally reinforced rather than challenged colonial logics. This is not to say that migrants' ethnic ties do not remain significant in enabling the construction of cross-border networks; it *is* to say that they may not necessarily involve emotive ties to a real or imagined "homeland," or mutual self-help, as in the traditional model. While the concept of diaspora clearly remains relevant as a vector for transnational organization, it must be problematized and placed in context. The Taiwanese case may, therefore, be helpful regarding the role diaspora plays in transnational ethnic communities.

Given the longevity and visibility of the Chinese diaspora, it is unsurprising that it has been one of the most frequently described in the literature (see Nonini & Ong, 1997; Cheung, 2009). In particular, studies of skilled labour migration have focused on how ethnic, family, and other ties to China serve to develop profitable business networks (Dahles, 2005; Tien & Leung, 2015). This often includes particular reference to the concept of *guanxi*, a difficult-to-translate term that Yang (1994) defines as "dyadic relationships that are based implicitly (rather than explicitly) on mutual interest and benefit," focused on the mutual exchange of favours over time (qtd. in Xin & Pearce, 1996, p. 1642). Many books and articles have been written about *guanxi* networks, for instance, about how they are used to acquire useful ties on the mainland (Chiu, Wu, Zhuang, & Hsu, 2009) and to further transnational business activities, both legitimate (Tien & Leung, 2015) and illegitimate (Gillespie & McBride, 2013), as well as for networking within organizations (Park & Luo, 2001). Also, it has been documented how Taiwanese skilled labour migrants and business people use *guanxi* for networking with one another and with mainlanders (Chiu, Wu, Zhuang, & Hsu, 2009; Deng, Huang, Carraher, & Duan, 2009; Numazaki, 1986). *Guanxi* networking is arguably not unique to the Chinese diaspora, given that many diasporas develop similar types of connections (see Xin and Pearce, 1996, p. 1642; Alston, 1989). Nonetheless, the literature on *guanxi* sometimes verges on the orientalist, treating it as something unique to the Chinese that outsiders struggle to understand and/or access rather than as an example of a more general transnational social activity related to similar practices around the world, which may be more or less familiar to non-Chinese depending on one's personal and/or cultural background. The relevance, uniqueness, and exclusivity of *guanxi* should therefore be questioned.

Similarly, the concept of the "bamboo network" – that is, family-based networks of Chinese migrants – presupposes that Chinese connection-building is somehow qualitatively different from what is found among other migrant groups, and presumably more effective. This idea is challenged by Tien and Luan (2015), whose study of Taiwanese cross-border networking found that the "bamboo network" had little effect in practice. Again, it is worth asking whether or not Chinese skilled labour migrants are different in this regard from those whose networking is less often studied, such as the French (Ryan & Mulholland, 2014), the Germans (Moore, 2005), or the British (Beaverstock, 2011, 2018). Furthermore, the concept of an ethnic diaspora–based network ignores the fact that, as seen in previous chapters, identity is multivalent and flexible; a supposedly "Chinese" ethnic network is going to be complex and disputed within itself, for it will include those Taiwanese who, entirely or situationally, may self-identify as Chinese, Malaysian Chinese, American-, Canadian-, and Australian-born Chinese, Hong Kong Chinese, and many others, as well as mainland Chinese (and the mainland itself is riven with ethnic, cultural, and linguistic divisions).

Finally, it seems likely, given what we saw in the previous chapter regarding the flexibility of identity and how Taiwanese identity varies with context, that a supposedly monoethnic network is likely to allow connections with other ethnic groups, using, for instance, discourses of pan-Asian identity or links with host-country populations.

Family on the Mainland: Taiwanese Connections with China

Guanxi, Taiwan, and the Mainland

It has been a common trope in the literature on Taiwan and globalization to consider the country in terms of its social and historic connections to mainland China. Studies of Taiwanese transnational businesses and expatriates, for instance, often focus on their ability to network with mainland partners above all else (e.g., Deng, Huang, Carraher, & Duan, 2009). This was acknowledged by some of my informants: "China became powerful and then Taiwan, we use our special, very unique advantage: the same culture, the same language [as the mainland] and we can communicate easily from coast to the inland of the China" (Steven). Mr Lee, after telling me a slightly sanitized version of the story of the post-1949 migration (which, significantly, given that he was affiliated with the Taipei Representative Office, conflated Taiwanese identity with that of the *waishengren* incomers), indicated that many people had strong social ties with the mainland: "We still have the connection with other [Chinese] people, just in business it doesn't matter, but in culture, in loyalty, because most of people move out their roots are in China." It is no surprise that descriptions of Taiwanese transnational networks offered by civil servants and state representatives reflected the literature on networking in the Chinese diaspora as well as the official KMT line on integration with the mainland.

In Taipei, similarly, Frankie, a local area manager for a European automobile MNC, described a fairly classic progression of *guanxi* networking with business partners in China, which involved developing connections over time through the repeated mutual exchange of favours:

> When I develop a business in China, maybe ten years ago, I only visit the customer twice I think. The first one is ten-thirty, morning, okay, go to visit customer, just handshaking, something, then we go to the restaurant for lunch, okay, then drink the wine, something, yeah, then everything's okay. The next time you come you've got a relationship, then the second visit, four-thirty, also for dinner you know, so everything is okay, just eating, then drinking, yeah ... so must become the party-party first, it's very important.

My study did, therefore, show that there were indeed active social connections with mainland China, and this could be interpreted as conforming to the impressions given by the literature on bamboo networks and *guanxi*.

Mainland China was not, however, viewed as a desirable network connection by Taiwanese skilled labour migrants. In their narratives, they often portrayed the mainland as highly repressive, as well as strongly localized and focused above all on money. Lucy described Taiwan as a "very fusion type of place," contrasting it with the mainland: "In China, if you don't do the old way, people are like, 'You have to do it the traditional way,' but in Taiwan, people don't care about it as much." Several Taipei interviewees described mainland Chinese as focused solely on material wealth and monetary gain. One interviewee, an art dealer, said that the mainland Chinese were primarily interested in the money value of a painting rather than its aesthetics, and Marina, an online community manager for a gaming company, said that the Chinese only did business with Taiwanese because they perceived the Taiwanese as subservient and obedient. The implication here is that in practice, ethnic and *guanxi* networks are complicated and far from harmonious, as well as shot through with discourses of power and colonial relations, and are hardly simple mutual-aid networks built on ethnic and family lines.

Mainland Ties in the Diaspora

In London and Toronto, not surprisingly, I also found that even informants who were *waishengren*, and/or who were generally positively disposed toward the mainland, tended to be fairly cautious and ambivalent when discussing it:

> At the moment Taiwan relies on China too much. Yeah, like, I would never put all my business on the Chinese, China market, yeah so that's the problem ... Taiwan can do more with European or American or British [businesses], or anyone else, yeah. It's really, it's nothing wrong to do business with China, but just really as I say, yeah, all of their eggs in one basket, that's no good. (Maria)

When positive aspects of working with the mainland were raised, many, such as Mr Lee, discussed this in rather instrumental terms:

> They [Taiwanese businesses] now go to China, where a lot of them spend the investment there, especially after Tiananmen Square. All the Western country they move out, and they take the advantage in moving, but at that time some of the people, they say that China is no good, so the [Taiwanese] Government forbids it, but the private sector, they still go there. So first go, first win. The other ones just follow the Government, the restriction law, then they go, maybe, I think, in the past two or three years, couple of years, then the Government lifts the ban officially, allows the people to invest in China, that's the whole situation.

Here, Mr Lee is describing Taiwanese businesses as going to China to seek economic opportunities abandoned by Western businesses rather than as motivated by more positive social ties. It is significant that in doing so, he references the Tiananmen Square massacre, a symbol of Chinese oppression. One of the business networking associations involved in this study had, in fact, been opposed to reunification with China in the past; although it was, according to rumours in the local Taiwanese community, changing its stance in light of the mainland's current economic power, its members were presumably still experiencing the same ambivalence toward it as before. *Guanxi* was not seen in terms of positive ties to the mainland – as a fount of home, family, and diaspora connections. Rather, networking activities with mainlanders and mainland businesses were viewed as instrumental, contingent mainly on China's economic power, and Chinese partners were not necessarily more desirable than others simply because of the diasporic connection.

Kevin, as a businessman, was more critical than most of China as a partner:

> The business we had ... I would say you know, [was] initially Taiwanese, and then they all move their factories to China so all the export to here, all from China, Hong Kong, but that part gradually faded because most of the Taiwanese companies, they have subsidiary companies in Europe, either in Holland or in UK or in Germany ... as the headquarters of warehouse, and they distribute from there to Europe, to East[ern] Europe, to the UK, to everywhere, so gradually I don't need to import from the Far East, and I still have a concern, buy from China directly – in fact, I never done that, I don't want it, because I feel Chinese and Taiwanese, Chinese and Taiwanese different, entirely, very different, I can't trust them, I can't trust them.

That extract calls for some unpacking in terms of the Taiwanese migrant connection with mainland China. Kevin was someone with mainland connections, who was succeeding in doing business with the mainland, yet he emphasized that his decision to do so had been instrumental and that he would happily change course when the opportunity presented itself. Other London interviewees also implied – and some declared outright – that the mainland Chinese had a reputation for shady dealings. While Kevin does appear to be maintaining some form of diasporic and/or *guanxi* networking, in that he prefers to do business with Taiwanese companies with European subsidiaries rather than to work with European suppliers – indeed, he founded his business by using his network connections in Taiwan to buy and import Taiwanese electronic goods to the UK – the idea that Taiwanese identity essentially entails positive relationships with the mainland, and the construction of ethnic networks with mainlanders, seemed to be less the case than expected. The reality seems to reflect more ambivalent connections with the mainland focused mainly on economic opportunities at the time.

Mainland Connections in Taipei

In Taipei, networking with the mainland seemed to involve a degree of ambivalence. Frankie, for instance, who worked for a European MNC with operations in Taiwan and on the mainland, hinted that such relationships could often exist at the expense of quality:

> Yes, yes, of course in China it's also the relationship but a different kind of relationship. [*Interviewer: Are we talking about guanxi?*] Yes, yes, it's very important in Chinese social relationships, yeah, it's very important, to win

> the project you must have a good connection, yeah ... Japanese people are looking for what kind of expertise you have, but most of the Chinese people is the buddy-buddy relationship ... Yeah, "you are my buddy, okay, your product is okay, I can buy your business," but in Japan it must be – the product quality must be okay, very good, and then you can make the business ... In China it's the relationship first, or the product maybe sometimes is okay, but you also can get business, so you know, in China most of the time your business is [a] discussion round the table – [but] not an office table, it's a restaurant.

In other words, Frankie felt that the most important aspect of business for the Chinese was the relationship with the local partner rather than the quality of the product being produced. So while *guanxi* and ethnically based networking were practised with the mainland, it was usually viewed at least with ambivalence and could even be seen as something not particularly desirable at all.

This links into a Taiwanese local aspect of identity, that of having "preserved" China's pre-Communist high culture. After I watched a Taoist parade go by my hotel and described it on my Facebook wall, Gordon, a Canadian friend who had been an expatriate in Taiwan, responded:

> So glad you got a look at one of these! It's all rather like the crowd on the bridge going into the bath house in [the film] "Spirited Away" or a low-tech version of the parade in the second "Ghost in the Shell" movie. It's one of the big differences between Taiwan and the Mainland. When I got back to the office with all my snaps of one of these things several of the Chinese-Canadians in the office with a Mainland ancestry were astounded. They had no clue as to who all the spirits were and said that maybe their grandparents might know. It kind of looked like an SCA [Society for Creative Anachronism, an historical reenactment society] event to them!

This sort of practice could, therefore, be seen as a way for Taiwanese to differentiate themselves from the mainland, that is, by maintaining and preserving traditions that mainlanders (and at least some diaspora Chinese) considered old-fashioned and quaint. Yet other informants, particularly in Taipei, told me that in recent years, the mainland Chinese had themselves shown interest in reviving these traditions and learning more about the Imperial past. One of my interviewees in Taipei was involved in designing and arranging a tour of treasures from the Imperial Museum to mainland China, which she saw as a means of fostering rapprochement between the two cultures. The association of Taiwanese identity with a kind of aesthetic, romanticized Chineseness continued even in contexts where Taiwan was no longer generally seen as the locus of "pure Chinese" traditions. Furthermore, this association could be simultaneously a source of conflict and a point of commonality with mainland Chinese.

One significant aspect of Taiwanese networking with mainland China initially emerged in the UK cohort and was repeated later by the Taipei cohort. This was the idea that European businesses sought to work with Taiwanese businesses not simply because they regarded them as good partners, but also in order to develop mainland connections without risking the reputational damage of being associated with mainland Chinese labour practices. This was described laid out for me, somewhat diplomatically (and using some allusions to kinship networks), by an official of a Taiwanese trade organization:

> The reason is because our product, our economic situation, we have to make differences because the Chinese right now, their image is not good. So that's why we, Taiwan, have a good image. But we think, we believe Taiwan has this good image because of such kind of a situation, it is also good for China, for Chinese, because that give[s] them a goal to approach because they also have same brother or same Asian people, same people, okay. But Taiwan having a good image is also good for [the] Chinese situation ... They can approach with Taiwan ... *[Interviewer: So they can use Taiwan's better reputation on the international scale as a kind of way of getting themselves accepted and working with Taiwanese partners is one way of kind of looking more respectable?]* Yes. Yeah. Because [speaking for] Taiwan, we hope to use this market as our background. (Dougal)

Something similar was later referred to in more blunt terms by Frankie in Taipei: "Some of the German companies, they will pursue the Taiwan[ese] companies to do business, maybe some kind of joint venture or for set-up of the other company to do the business inside China, because we and the Chinese people have a good relationship right now, so it's easy to do some promotion job inside of China."

This presents us with an image of European businesses using Taiwan as a kind of reputational stalking-horse for business activities in China; conversely, the Chinese will circumvent their own difficulties when conducting business with European companies by resorting to Taiwanese middlemen. Here, then, the Taiwanese benefit from the perception that they have close network ties with the mainland. Yet the value of those ties is somewhat morally ambiguous and, again, does not suggest the development of positive relationships with the mainland.

While this study did confirm that ties to the mainland were important for Taiwanese business people both in Taipei and in the diaspora, responses also pointed to a fair degree of ambivalence regarding whether those ties were generally a positive thing. This aspect of networking has not really emerged in other studies of Chinese networks.

A Global China? Taiwan in the Chinese Diaspora

Diaspora Orientations in London

It might be more useful to see the Chinese aspect of Taiwanese identity as relating not to the mainland but rather to a wider identity common to all ethnic Chinese, worldwide. Yu and Bairner (2008) note that the KMT, in the 1970s, looked to the Chinese diaspora to support its claim to be the "true China," and for *benshengren* interviewees such as Wendy, identifying as part of the wider diaspora was a way of expressing an alternative, non-mainland, Chineseness:

> I made a lot of Chinese friends or students while I was [at university] in Manchester, I need to provide an example … [One friend of mine,] he's a bit different, he's German-Chinese, so basically, he's in between … I would say, for example, we still feel the connection between Taiwanese people and people from Hong Kong is closer than Taiwanese people to Chinese people, even we know Hong Kong is a part of China, but they still consider they are different. (Wendy)

Wendy also indicated that Taiwanese students, while they could make use of Chinese associations, often did not do so:

> [Universities in] the UK have Taiwanese societ[ies], so we join [the] Taiwanese society. We won't join some other, just maybe Chinese societies, and even [then] we all speak Mandarin, I mean, but we still join Taiwanese society. And for me, it's just like, in Manchester, usually I met Taiwan[ese] people there, and for this reason, I remember, later, I can work for whatever Taiwanese people, if we walk down the street, we see people from Asia, if we don't know them, definitely they are not Taiwan … But I would say, for example, we still feel the connection between Taiwanese people and people from Hong Kong is closer than Taiwanese people to Chinese people, even we know Hong Kong is a part of China, but they still consider [that] they are different.

Significantly, she repeatedly told me that she was not necessarily unhappy with being associated with Chineseness, but that she preferred to identify with diaspora Chineseness rather than associate herself, and

Taiwan, with the mainland. In London, groups such as Xinbeitou 8 are not simply places to connect with other Taiwanese, since they include anyone else who wants to speak Mandarin or, indeed, who is interested in language learning more generally. In practice, therefore, the group is organized around a global Sinophilia rather than a sense of geographically located Chinese identity, in that it welcomes anyone who wants to practise Mandarin (or some other Chinese dialect) and is willing to acknowledge Taiwan as a legitimate sociopolitical entity in its own right.

There was also the somewhat extreme case of Rupert, whose identity as Taiwanese was very much based on membership in the Chinese diaspora and on political affiliation, rather than on place of birth:

> My situation is even more complicated because I was born in Shanghai. And I left China in 1949, when the Communists came over to China and I went to Singapore. And when I was in Singapore I was given a Singaporean passport. But I never regarded myself as Singaporean but I regard myself as Chinese. But I cannot ... You know, I cannot accept the Communist Chinese is a free Chinese, you know. So, I joined the Kuomintang in London. You know, when I became the head of the Kuomintang in London and I started the [British] Taiwanese Chamber of Commerce. So, you know, my background is even more unusual in the sense that I was not originally indigenous Taiwanese. But because I show my allegiance to the Republic of China, I actually took out a Taiwanese passport and gave up my Singapore passport.

It should also be noted that Rupert had spent most of his life in London and was involved more with global organizations than with local ones. Furthermore, he was well-known and generally respected within the Taiwanese community, and while people did challenge the KMT's dominance of Taiwanese identity discourses, nobody, at least in my hearing, challenged Rupert's self-identification as Taiwanese. This allowed his political identity to be read back to him as an ethnic identity. Rupert's rejection of Singapore is not insignificant to his *waishengren* take on Taiwanese identity, for Singapore is a good example of a Southeast Asian country that is dominated by a large Chinese population but that nonetheless does not identify as a Chinese nation: by explicitly choosing to live in "Free China," Rupert is suggesting that identification as Taiwanese can be a means of choosing how one interprets a diaspora Chinese identity. Taiwanese identity is thus reconstructed, through the complications of post-1949 politics, so as to connect to a global Chinese identity focused on ethnicity and diaspora, and different ways of embodying these, rather than exclusively in terms of relationship to the mainland.

Diaspora Orientations in Toronto

Compared to the British cohort, the Canadian one tended to describe their connections with China even less in terms of the mainland. Instead, they tended to talk about their relationships with other groups of diaspora Chinese in Canada: with the local Cantonese community, which was described as dominating the community, and which viewed the Taiwanese as rivals; and with the mainlanders, who were viewed as recent "incomers" often ignorant of Chinese Canadian ways. Lucille, a real estate agent, describes Chineseness in the Canadian context very much in terms of waves of migration rather than in terms of mainland and Taiwan:

> Before 1997 [when Hong Kong rejoined the mainland] ... you can say that was a time that the [Cantonese] people came to Canada because they weren't sure how, after 1997, other societies are going to be, you know, up to change, routine change, so they were, and then in real estate time ... at that time there was a boom, so you know, [the Chinese community in the Toronto suburb of] Richmond Hill almost probably was at that time established. Then up to 1997 or 2000, probably up to a few years before that, it's come down a bit, more stable, I don't know, but for some reason people felt that Hong Kong has more opportunities than here, so they were kind of, another time, [inclined] to go back. And that was up to then to now ... We see more from Mainland China these years.

Nora, like Wendy in London, described being Taiwanese in Toronto as very much a transnational identity, embedded in a global diaspora, while also making reference to her more local identity as someone from Chentso village:

> After I come here, I might say that the biggest thing I learn isn't that Western civilization, like I thought, the biggest thing for me to learn in my life is that I knew those Chinese people, all over the world, when they see a Chinese people, Taiwanese Chinese people here, Mainland China and Chinese people here and it's so different, and Hong Kong people, they are so different, so maybe next time you can [study] Hong Kong people here, yeah [laugh], and Malaysia people here. Or maybe you can do like the comparison of Chentso people here, and the Taiwanese people here because [of] their ancestors [laugh].

This suggests that before leaving Taiwan, she had not necessarily thought of her identity in terms of the Chinese diaspora, but that the immigrant experience shaped her view of Chineseness as a global rather than a national identity.

Taipei and the Issue of Location

In Taipei, Teresa made an observation during an interview that initially seemed to relate to Taiwanese–mainland relations but then turned out to relate to the Chinese diaspora more generally (since most people in Taiwan did not consider people from Hong Kong to be precisely the same as mainlanders): "Because my major business is still focusing in the greater China area, I find it really easy being a Taiwanese, [I can] deal with people from Hong Kong, from China, it's very easy to deal with them with this identity." This sums up the way in which the expression of Chineseness as an identity could be complicated and relate to both the mainland and the diaspora on different levels, within Taiwan as well as outside it.

A remark by Nora (Canadian cohort) seemed to indicate that Chineseness could be influenced by location:

> My husband is more like Taiwanese Taiwanese [meaning *benshengren*] so when we got married, if he is not the youngest son and his mother died and it was in a hurry, maybe my husband's family would not like a *waishengren* daughter to go inside their family ... I think you chose a very good topic. Because Taiwanese here, that's Chinese here, because Chinese from Mainland China Chinese, they are the same, but Taiwanese they separated; two different, like some of them are very very older Chinese [i.e., have the culture of ancient China] ... Or like Johnny you interviewed, he is more *waishengren.*

In this excerpt, she first alludes to the social divisions in Taiwanese identity, between pre- and post-1949 Taiwanese, but then indicates that because Canadians generally perceive her as Chinese ("Taiwanese here, that's Chinese here"), and because mainland Chinese identity is itself regionally differentiated, the Taiwanese are in a complex situation – simultaneously perceived as part of a "Chinese" identity yet with that same identity split into regional identities.

Lucille viewed this complexity of identification as part of the Taiwanese experience, inherent in its colonial history:

> Because I'm always very mixed ... All the friends I had [at school] are more Mainland Chinese in that sense, at that time. So there are many different type of people, right, so I thought we have a little bit of influence by the

> parents, educated under the Japanese occupation, and that sort of culture [is] more rigid, self-desecrate, very, and take responsibility and all that. [Pause]. But we have good friends as Chinese, but some impression about the Chinese would be ... [laughs], Like I say I really don't know how to use that word.

This presents Taiwanese identity as connected with other, local and global, Chinese identities, but at the same time as complicated by Taiwan's connections elsewhere. Identifying as part of the Chinese diaspora is, therefore, down to individual self-presentation but also influenced by the social environment of their location and by discourses of what it is to be "Taiwanese" in Taiwan and in the wider diaspora.

Asian Tiger: Taiwanese Identity as Pan-Asian

Pan-Asian Connections

The "pan-Asian" identity discourse discussed elsewhere in this volume offered another means of presenting Taiwanese identity, one that served as an alternative to the "Chinese" discourse by recontextualizing Taiwan as part of another form of transnational ethnic or "meta-ethnic" network. The development of a pan-Asian transnational identity through shared symbols such as particular brands has been observed in the marketing field (see Cayla & Eckhardt, 2008). For the Taiwanese, however, it provided an alternative take on their own identity, one that could include the Chinese aspect without making it the central focus.

The development of a pan-Asian identity for Taiwanese was evident in my interviews. Many interviewees, in all locations, would, when asked about Taiwanese identity, told a very similar narrative relating to Taiwan's historical role as a seafaring, island culture that had built connections throughout Asia:

> We have a tradition, especially for those people live in the boundary area in the sea coast, of course they go by sea and go down to the South Asia. For those living in the Western area then they go Taipei then they go to India. Some go to the West, go through to Russia and other places, Saudi Arabia or even to the Europe. (George, London)

> Start from seventeenth century, Taiwan was actually quite cosmopolitan place because at that time you have Portuguese, we have Spanish and then Dutch. But then being a Chinese when I was young, well as a Taiwanese I was taught that I'm a Chinese, and then, but then when I started to work

> into modern society it most of a sudden change and people telling that we are Taiwanese. And by doing this [running a digital animation and design company] for me is really to search for my own identity. (Ms Wong, Taipei)

> Taiwan is an island ... so trading is very important in Taiwan, so Taiwanese people it's contact with the outside before the Chinese people ... Most of the people, they are thinking how to get a business inside [their country], but in Taiwan the people are thinking how to win the business outside [their country]. (Frankie, Taipei)

Some, however, like Nora, constructed this more negatively, suggesting that it is not Taiwanese culture itself that is transnational, but its migrants: "I always think that Taiwan is such a small island, our life is very short, why not live in another country for a period of time, right, longer time." However, her remark also suggests that for the Taiwanese, international travel and building connections is a corollary of their smallness and island status, again suggesting an identity tied up with transnational connections.

In this context, present-day investment in other Asian countries besides China also was emphasized as a marker of Taiwanese pan-Asian connections: "Taiwanese invest[ment] in Vietnam is number one, also that in Thailand, Malaysia, Indonesia, and you can use the Taiwanese business community in those Asian countr[ies] to invest or to explore those markets, I think it's very good" (Steven).

Sarah, working in Taipei for a global search engine company of American origin, described a global mindset as one of Taiwan's selling points for investors: "Taiwanese have an interesting situation, they like to listen and learn from the global. In one side they were afraid you control me, but they always like to listen and learn from the global because they think global business is more, is more, more, more bigger, more affluent that local, they all think that, but they're always scared, they have a complicated mindset."

However, she also saw having a "global mindset" as something that was focused on things other than profit and business opportunity, and suggested that Taiwan's concept of itself as a global identity might be rather limited:

> I think that Singapore is more global, global mindset you know, because every time I discuss business with the Singapore colleagues, they were thinking about them a while, and they were thinking about more, more, more open, they won't think that you want to control me, no, they will think how we can make more opportunity, more possibility or how we win from both sides not just me, I think it is the big difference, Taiwanese people always think how we can, how can I win, but I don't care, I just, this is the biggest difference I think.

Although the Taiwanese would, to some extent, identify as "pan-Asian," or draw attention to Taiwan's wider Asian connections, as an alternative to discourses related to Taiwan's Chineseness, these did not seem to relate to building connections and developing social networks in the same way as the Chinese identity discourses above.

Like the Chineseness discourses, the pan-Asian discourse reflected a certain ambivalence, largely linked to Taiwan's cultural history and past relations with other Asian nations:

> [My research assistant] asks me about Japan, what it's like. She's never been there but says a lot of Taiwan looks to Japan, in terms of fashion, business etc. It is also increasingly looking to Korea, but she says a lot of the Taiwanese don't like the Koreans. I ask why and she thinks and says she thinks it is because they are always claiming things for themselves, e.g. they claim Confucius was Korean. (fieldnotes extract, 9 April 2013)

Glinda, in an interview, said she thought that younger Taiwanese were very Korea-focused, due to the rise of "K-pop" music as a regional phenomenon, but added that she herself did not find Korean culture appealing and much preferred the Japanese. The romanticization of the Japanese occupation period by my Taipei informants, discussed earlier, also serves as a symbol of Taiwan's historic connections with other Asian countries, and subtly underscores that these can be more positive than those with the mainland. It therefore seems as if the concepts of diaspora, *guanxi*, and the bamboo network, as well as Taiwan's place in Asia more generally, remain a crucial basis for networking, and for developing a global identity, but are all much more ambivalent and politically charged in practice than in theory.

Connections with Host Societies

It should come as no surprise that all of these ethnic networks were clearly affected by the social context of the host society. In the UK, for instance, the role of the Taipei Representative Office was seen in terms of developing connections with the British; however, the nature of the diaspora, and the political affiliation of the representative office (which at the time supported reunification with the mainland), meant that connections with the wider Chinese community were made instead, albeit on a temporary, politically motivated basis: "Okay, my position is for Taipei to be here to help our compatriots in Britain, to help them to organize some society, to organize some activity, to have friendly communication more often, to help them have a better connection with others, yeah. In the past mostly our help is more broadly definitely Chinese but for the past three years there is more emphasis on the Taiwanese compatriots" (Mr Hsiao, Taipei Representative Office).

Michael, here, discusses how the networking roles of chambers of commerce are affected by the culture in which they find themselves and how those roles extend out to other diasporas than the Taiwanese and the Chinese (besides, incidentally, including alumni and professional identities):

> Of course I want to get our Junior Chapter not only with the Taiwanese people but also with Japan, European, German, French, and also the British people ... Sometime we engage to like our Junior Chapter and Cambridge, Cambridge Society for the young students, for a young person. The social networking will be getting bigger and bigger and also we cannot always stay in the talent circle, we need to get more connection from international. (Michael)

Nora, in Toronto, indicated that local organizations shaped ethnic identification. For instance, she joined an association for people with ancestors in a particular town, Chentso. But later on, that group drew members from all over the diaspora:

> So I joined the Chentso Association of Toronto, and they're the group of people, the association, mainly they are from China, right, but they invite us that are Taiwanese, [all] those people whose ancestors are from Chentso ... Because of my daughter's performance I knew them, and then I joined them and they loved what I did because I can do the program in the annual of big party, so I become very important, the director of the board of directors of our association.

The fact that in Toronto there were local groups for people from particular towns or regions could shape the expression of Chinese, and Taiwanese, identity and the very concept of what constitutes identity in the Chinese and Taiwanese diasporas. An ethnic identity can also include, and be included within, a regional one.

Lucille, similarly, indicated that one's individual sense of Chineseness could be affected by the nature of the Chinese diaspora. She alluded to the changing demographics of the Chinese community in Toronto:

> At one time that I went to library and I saw those [Cantonese books] labelled "Chinese." I told the librarian, "They are not Chinese as such, you know, it's not [the] official language." She replied to me that "how will I know" [laughs]. So they didn't really care. But gradually I think probably

> in the eighties a lot of Taiwanese came, particular in that area, then it was recently, probably up till 1995 that period of time, up to now it's more Mainland China.

The incident with the librarian reflects the rivalry the Taiwanese feel toward the Cantonese as well as the fact that non-Chinese Canadians generally perceive Cantonese identity as normative. It also indicates that Taiwanese people wanted their distinctive take on Chineseness to be recognized by outsiders. The rise of mainland China, where Mandarin is the official language, to prominence had boosted Taiwanese visibility to some extent, in that Mandarin was now recognized more frequently as the normative Chinese language – although this might well bring the additional problem of Taiwanese diaspora members being mistaken for mainlanders). Ethnic identity is thus subject to local discourses as well as to home-country ones. It is also a dynamic entity, changing as the population changes.

Many of my interviewees had widespread family ties, suggesting the "bamboo network" of the literature, and suggesting as well that they were part of a locally rooted but globally connected ethnic group. Vivian, for instance, who had lived in various places in North America before eventually settling in California, had a typically scattered family:

> We came to Taiwan when I was six years old, in 1949. My father worked for the government so we were able to escape the communists ... My parents have passed away but they lived in the States when they retired. And I have three sisters and one brother, one sister in Chicago and one sister and one brother in LA ... And my brother, he is blind, but he is an amazing story, he does lots of church work and even open schools in China, mainland China. He has the money to do charity work, God's work, ministry.

One other point to note in Vivian's account is that, like Penelope's work with refugees and migrants on behalf of a Buddhist organization, it shows that religion can generate forms of transnational networking. In Vivian's case, the evangelical Christian church (to which she and her family were converts) led them to engage in missionary activity. The obligation to evangelize that Vivian's family drew from their faith meant they maintained global ties that were not necessarily ethnic but could combine with ethnic and diasporic connections. While this was generally more significant for older informants (although one younger informant had converted to evangelical Christianity while studying in the UK), it is still worth noting as evidence of how constructing nominally ethnic transnational networks in fact connects to a variety of symbols of identity.

The use of ethnic connections to form social and professional networks was, therefore, affected not only by home-country discourses and personal affiliations but also by host-country dynamics and a variety of other nominally external factors. The settler discourse of the Chinese community in Toronto, for instance, means that Chineseness manifests itself as part of a wider ethnic community that is divided according to more specific origins and dialects, whereas in London, the issue was driven more by domestic Taiwanese policy. Furthermore, the use of ethnicity as a basis for networking was complicated by other identities, personal and professional. This was due to the multivalency of symbols, meaning that the same symbol can be used to present different identities simultaneously.

What's in a Network? Discussion and Analysis

Diaspora Identities and Cross-Border Networking

In some ways, this study supports the literature suggesting that ethnic identity is a crucial way of operating across borders in the network society, particularly for Chinese migrants. Diasporic connections clearly existed for my informants, as did more abstract ties, whether to the mainland or to a sense of global identity as "Chinese." However, rather than being a straightforward set of networks based on shared ethnicity, these were complicated by other factors, with mainland connections generally being a more immediate aspect of Chineseness for some *waishengren*, and *benshengren* often focusing on Chineseness as a global identity shared with people who were not necessarily mainlanders, such as Malaysian Chinese or Canadian-born Chinese. For the Taiwanese, ethnic identity was complicated by internal and external differences in interpretation and was frequently questioned, reinterpreted, or problematized.

Furthermore, as per Ho's (2011a) argument, in the Taiwanese case ethnic identity was also open to co-optation or exploitation by external entities, such as the government. Mr Lee, from the Culture Section, hinted strongly at this early in my interview with him:

> Yeah, [Taiwan is] the Republic of China, so we think we are China. So at that time we take care of the people, all those Chinese people stay[ing] overseas, and then the Governments there take care of them. So we have a special department, the ministry level, Overseas Chinese Affairs Commission … Even before 1949 and afterwards even the Government, Central Government moved to Taipei, and that is the special department at the connection with all the people around the world, South America, Europe, North America and South Asia, everywhere.

Effectively, then, the historic KMT position was that it bore responsibility for diaspora Chinese, as part of its rivalry with the mainland. This stance continues to inform Taiwanese policy toward diaspora Chinese in the present.

The Taiwanese student organizations, furthermore, being government-sponsored, could similarly be seen as an official means to exploit ethnic identity to further political aims:

> The Government requires us to balance the trade, especially the education service area. We ... encourage most [Taiwanese] students to come, but we also like to have some foreign students to go to Taiwan ... They have the international environment, it help them to practise their language, to have broadened views, so that's the other way we want to balance the trade, balance the culture, to bring more students to go to Taiwan. And we try to recruit some English teachers to teach in Taiwan for the primary schools, for junior high schools, but we also like to have some professors to teach in Taiwan. (Steven)

The encouragement of networking with the mainland, furthermore, supported the KMT's aim of a KMT-dominated reunification of China, since it acknowledged the ties between the countries without endorsing the Communist claim to the mainland, or to Chineseness as a whole. This provided another driver for politicians to encourage the perception that Taiwan was part of a wider Chinese diaspora. Discourses about transnational entities therefore could be, and frequently were, co-opted for political purposes.

However, such discourses were complicated by individual actions and preferences. As discussed above, most of my interviewees were ambivalent about connecting with the mainland, even as they noted the shared ethnic identity, and instead often focused on Chineseness as a global identity, which rendered the situation much more complicated than simply a question of whether an official policy of communist or capitalist ideology should be associated with Chineseness – since, if connections could also be built with overseas Chinese and Hong Kong Chinese, then a variety of discourses of Chineseness could coexist within the global identity, and China as an ethnic identification would be bigger than simply Taiwan and the mainland. Although, arguably, the discourse of a "pan-Asian" identity might seem to be a weaker one than the idea of an ethnic diaspora, it also was compatible with the idea of Chinese identity as global, for it encouraged an identification with Chinese populations elsewhere in Asia.

Furthermore, symbols could be open to multiple interpretations. For instance, although the student organizations were supported by the

Taiwanese government with the aim of furthering its particular take on Taiwanese identity, the students who participated in them did so for their own ends, and the same organizations also wound up providing support for professional networking, through connections with alumni associations and other forms of ethnic networking. Likewise, ostensibly ethnic organizations generally turned out to be multipurpose in practice, with people using them to network for their own purposes. The agency of individuals, and the ways in which they use and construct diaspora identities, is clearly a key factor in how ethnic and other networks are used.

Although Chineseness is a source of identification and networking for community members, it is more complicated than simply links to Taiwan, or between Taiwan and the mainland – connections extend all over the world. Furthermore, people engage with their Chineseness through organizations that are not exclusively ethnic – whose focus is on business, religion, academia, or gender. Even the more obviously diasporic organizations and events, such as Xinbeitou 8 and the food festivals, are not exclusively Taiwanese. Furthermore, the question of whether one identifies as Taiwanese, Chinese, or diaspora Chinese is complicated, in that identities and affiliations easily become slippery, multivalent, and complicated by external power relations. The complexity of identity thus means that the diaspora networks not only extend into professional and other networks, and cross national borders, but also allow individuals to cross other borders at the same time. So the idea of an exclusive bamboo network proves more complicated in practice, and even the practice of developing *guanxi* is fraught with ambivalence and power relations.

The Importance of Context

Ethnic identity was also complicated by other identities. For instance, relationships with the wider Chinese diaspora were complicated not only by whether the speaker was *waishengren* or not, but also by whether that individual had become an adult before or after the late 1980s, which saw the end of the Cold War and, more locally, of the KMT dictatorship. This meant that Taiwanese, and Chinese, identity was filtered through different lenses: the Cold War cohort tended to think more about the national level, and raised different issues with transnationalism to the younger cohorts (as an extreme case, one interviewee in Toronto spoke of the danger of spies from both the KMT and Communist governments infiltrating the community, something that did not seem to be a concern for younger members). In many ways, the older generation seemed to conform more to the conventional image of a diaspora (see Cohen, 1997), of people migrating to escape a political regime or to pursue better economic opportunities. The younger generation were more like the globalized cosmopolitans of Iyer (2000) and Hannerz (1992, 1996; see Gonzales, 2019, for a revisiting of the concept), pursuing study, jobs, and/or relationships abroad. They were connected with the diaspora through relatives, but they also engaged more with global than with national social spaces. There were clear differences in the ways in which various groups related to the concept of the diaspora, which suggests that the same diaspora can be experienced quite differently by different members.

Furthermore, even among the older generation, patterns of migration could be complicated. For instance, Rupert's account (detailed earlier) of going from the mainland to Singapore and then deliberately choosing a Taiwanese identity as a political act, and that of Mrs Wang, whose family fled mainland China not because of any affiliation with the KMT, but because Mrs Wang's father was a businessman who did not want his company to be nationalized, call into question the idea that diaspora identities conform to a single general pattern. While Mrs Wang generally self-identified as Taiwanese, she would also self-identify as Chinese, depending on the circumstances – an indication that diaspora identities are complex and multivalent. Identifications, networks, and relationships with the diaspora also changed over time, in line with politics and changes in the local and global economics, as evidentd in the ambivalence felt toward doing business with mainland China expressed by many of my interviewees. Relationships with the diaspora were complicated, involving patterns of migration, further migration, and return migration, and with people identifying variously as Taiwanese or as Chinese depending on the circumstances, using the same symbols. This, again, suggests that we need to think in terms of diaspora less as a single phenomenon and more as a complex of identities and symbols relating loosely to ethnic origin and migration.

The Development of Global Chineseness

Another pattern worth noting is how Chinese identity, during this study, frequently became detached from particular geographical locations, instead becoming a kind of global identity focused on ethnicity rather than on birth or on a specific place of residence. While this could be for social reasons – for instance, to avoid the difficult question of whether Taiwan or mainland China had the greater claim to dominating the discourse of Chinese identity, or as a diplomatic way of dealing with the ignorance of host-country colleagues (who might or might not be aware of the nuances of Chinese versus Taiwanese identity) – the practice also seemed to be something more complex, pointing to a Chinese identity that was not necessarily identified with a particular geographic location. Glinda clarified this by explaining that there was a difference between China as a national entity, *Zhongguo*, and China as an ethnic identity, *Zhonghua* (see also Yu, 2011), adding that she, as a Taiwanese person, identified as the latter.

The presence of the Taiwanese diaspora, with its distinctive identity, as a subset of the Chinese diaspora, and of other distinctive groups that laid claim to a non-mainland but still-Chinese identity (such as the Malaysian Chinese or Hong Kong Chinese, or Nora's identity as a Chentso villager), highlights that there are different ways of being "Chinese" and of being simultaneously Chinese and something else. This produces a parallel to Björklund's (2003) observations about the Armenian diaspora: he argues that his informants do not view themselves as in a dyadic relationship with a real or imagined motherland, but "tend to insist on a meaningful identification with an Armenian 'nation' spanning many countries" (p. 337), and to the "Germanness as a state of mind" discourse that exists among German skilled labour migrants (Moore, 2005). So it might be useful to think of Chinese networks not in terms of a "standard" diaspora, and more as a kind of global identity spanning many different places; indeed, it might be more useful to see *many* supposed diasporas in this way.

In the context of the network society, identity is crucial in that it allows the construction of communities that may relate to location without being rooted in it, through the use of abstract symbols that can reference a geographical location without requiring one to have a specific connection to the actual place (see Moore, 2005). Chinese identity is not simply a matter of a conventionally defined diaspora or a particular set of network connections; rather, it is a global identity that spans multiple locations and includes a multitude of different and sometimes contradictory ways of constructing and expressing that identity.

The participants in this study, seen on one level, seemed to conform to the literature on bamboo and *guanxi* networks. They did use their Taiwanese and Chinese identities, and their social connections, to build ethnic networks and to participate in fairly conventional diaspora-building activities. Yet on other levels, they used the same symbols of identity to connect, through the diaspora, with other ethnic networks, and to connect with other networks that had little to do with ethnic origin. What could appear at first to be a simple, ethnically based transnational network on further inspection became complicated, leading into non-ethnic identities, with power relations between different groups coming into play, as well as different discourses providing other forms of identification, such as the "pan-Asian" view of Taiwanese identity. Individuals would identify as "Chinese," "Taiwanese," or something else, depending on the circumstances and the context. Furthermore, Chinese identity appeared to be increasingly seen as a global ethnic identity rather than one focused on a specific geographical location.

As with the question of host-country influence discussed in the previous chapter, membership in and use of the Taiwanese diaspora and the practice of networking through ethnic ties was complicated, as well as subject to continuously changing discourses through which identity was contested and redefined. The case of Taiwanese skilled labour migrants thus suggests there are crucial connections between seemingly localized identities and the development of non-local, transnational social forms, such as Taiwanese identity and global Chineseness.

Everyday Transnationalism: The Role of Location in Global Communities

The case of these Taiwanese skilled labour migrants, situated as they are in three different global cities, offers insights into the role of location in migration and the development of transnational networks. By considering the cohorts in their geographical nodes, we can explore how their experiences of migration varied depending on their experiences in Canada, the UK, and elsewhere, and consider the possibilities for differentiation and integration that the locations offer migrants. Furthermore, both of the overseas groups interpreted symbols of Taiwanese transnational identity in their own ways, meaning that their identity was the same as that observed in Taipei but also significantly different. Even though London, Toronto, and Taipei are all considered to be "global cities," the different opportunities they provide for the expression of identity, and the combination of ethnic and other identities in each location, make for different engagements with the local and the transnational.

Although writers on globalization and transnationalism have acknowledged since the early 1990s that "the global" and "the local" are inextricably entwined (for instance, Massey, 1991; Hirst & Thompson, 1996; Held, McGrew, Goldblatt, & Peratton, 1999; Tomlinson, 1999), only relatively recently have researchers begun to shift from treating transnational communities as if they were effectively unfettered in movement, to considering them as embedded and historically mediated (Ho, 2011b, p. 4; van Veen, 2018; Tseng, 2017; Lien, 2011). Much of the impetus for this comes from M. Smith's influential papers arguing for an "everyday transnationalism" approach (2001, 2010); he postulates that transnational migrants play a role in defining the identities of locations, through their movements and through their self-definition as a transnational community. Yeoh and Huang (2011), similarly, argue that rather than focusing on the hyper-mobility of transnational migrants, we need to consider their complex relationships with location and the place-making activities in which they take part (p. 682). However, the different impact that locations have on a single transnational group are only beginning to be explored, through, for instance, studies such as Teo's (2011) examination of Chinese return migrants, Tseng's (2017) consideration of Taiwanese and Chinese mixed couples and their migration decisions, and Ho's (2009; 2011a; 2011b) "trajectory-based" studies of Singaporean skilled labour migrants in London. By examining various different sections of a single migrant network in different locations, a study of Taiwanese professionals can shed light on the complex role of location in transnationalism.

In particular, while many studies of skilled labour migrants are conducted in global cities (e.g., Beaverstock, 1996; Dekker & Engberson, 2014; Moore, 2005, 2006; Ho, 2011a, 2011b; Beaverstock, 2011, 2018), and may consider the role that the transnational nature of a particular city plays in developing cross-border connections (e.g., Moore, 2005, ch. 4; Beaverstock, 2018), what is less often considered, in the context of transnational identity, is the fact that global cities are not all global in the same way (Bell & de Shalit, 2011; Acuto & Steele, 2013) or to the same extent (Parnreiter, 2013, pp. 21–24). More and more studies have indicated that "global cities" are hugely diverse (e.g., Bell & de Shalit, 2011). Krätke, for instance, identifies a global city network of manufacturing centres as a complement to the usual focus on advanced business services as a marker of globalization (2014), while Derudder and Taylor (2005) and Neal (2008) explore and evaluate hierarchies among global cities. Wang (2003) argues that Taipei can be viewed as a "regional global" city by virtue of being globally engaged in a Southeast Asian context. Finally, the migration of financial instituions away from the City of London following the UK referendum on leaving the European Union raises the possibility that a pre-eminent global city may become *less* global, and suggests a future in which Europe's financial, technological, and cultural industries are not concentrated in a single location but instead are distributed across several locations (Rubin, 2016). Clearly, then, globalization can vary over time in its nature and extent.

The identities linked with particular cities can, consequently, affect how transnational migrants relate to them. J. Waters (2006), for instance, explores how the specific cultural capital attached to a "Western education" in Hong Kong affects the drive among its young people to study in Canada. Massey (1991, p. 29) notes that places do not have single identities, indicating that cities can be local and global simultaneously. So it seems unproblematic to say that cities can be wholly or partly global in different ways.

Furthermore, the internal complexities of global cities are seldom discussed. For instance, while it can be argued that culturally, London has more in common with New York than with the rest of the UK (Derudder & Taylor, 2005), this is not true of all of the city's boroughs or populations, as Acuto (2013) notes in her exploration of how London's politics incorporate both global and local concerns. It is also worth considering whether there are differences in the ways in which Londoners and New Yorkers perceive, and use, their identity as people in global cities. The complexities of transnational urbanism, and global cities' connections with the ethnic identities of migrants, deserve exploration.

Questioning the Global City

Global Cities Compared

The first key point worth noting from the preceding chapters is that all three cities, while "global," were clearly "global" in different ways, and that these differences had a definite impact on the identity, self-presentation, and transnational connections of the study's participants. Ironically, although it is more common for London to named as an iconic global city than Toronto or Taipei, it is actually the one that *least* encourages immigration, instead focusing on a sojourner paradigm (with people assuming "sojourning" to be the default migration mode, whether it is or not). Likewise, Taipei may be the least visibly diverse city of these three, yet it maintains global connections through its unskilled migrant population, its diaspora connections, the IT industry, and the digital engagement of its citizens.

The experiences of each cohort also point to the idea that there are different globalizations rather than a single, emblematic globalization. The migrant experience is different in each city, and so too is those migrants' experiences of globalization and the network society. To illustrate this, note that in London, the norm for migrants was seen as working in advanced business services (ABS), whereas in Toronto, the norm, for Asian migrants at least, was seen as working in STEM fields; even though very few of my interviewees were actually working in STEM-related professions, several had qualified in STEM fields, and they often joked about how working in STEM was a "Chinese thing" or "Asian thing," which was not the case in London or Taipei (even though many more of my Taipei sample, relatively speaking, worked in computing and information technology). Some of this may relate to the settler/sojourner divide, with STEM providing arguably more physically grounded careers than ABS, but again this is complicated; transnational professionals exist in STEM fields, and settled ones in ABS. Arguably, it has more to do with the nature of globalization in both cities: where London specializes in finance, Toronto, although it has shown signs of transitioning to a more service-sector-driven economy since the 1990s, has been much more focused on industry and resource extraction in the past, and resource extraction industries still dominate the Canadian economy as a whole (Krätke, 2014; [Anonymous], 2014). The focus of both cities is global, but differences in the nature and experience of that globalization have had an affect on migrants' personal and professional experiences.

The dynamics and temporal nature of globalization also play a role. It is perhaps a banal observation that the reasons for migration among the Taiwanese varied depending on the time and the geopolitical situation: the older cohort's migration to both Toronto and London was influenced by Taiwan's position in the Cold War, as well as by the KMT's close relationship with the United States and its allies in the 1950s and 1960s (see Harrison, 2007), and their experience of Taiwanese identity was closely linked with the KMT and, during the 1950s and 1960s, with US efforts during the Cold War to encourage the perception that Taiwan was "free" or "pure" Chinese. Younger interviewees, having grown up in a post–Cold War world of transnational corporations, saw international migration and communication as a means to build their careers or develop their personal skills, in line with the "borderless world" ideal of the 1990s (see Iyer, 2000). But the fact remains that globalization varies not only from place to place but also from time to time, as does the experience of being a transnational migrant.

The Uses and Roles of Identity

The Taiwanese skilled labour migrants' experiences also illustrate how the flexible nature of identity can be harnessed to maintain a cohesive sense of belonging to a group even while adapting to local circumstances – in short, the networked existence proposed by Castells. The three cohorts maintained some recognizable commonalities: an interest in food, and in particular holidays and celebrations; the use of Mandarin, particularly when written in traditional characters; and a complicated self-definition relative to China and to the ethnically Chinese diaspora. The two migrant cohorts also linked Taiwanese identity with the idea of "pan-Asian" solidiarity, referring to Asian-wide foods, popular culture, and traditions (see Cayla & Eckhart, 2008).

Yet the same symbols also had different connotations in different places. In London, Hua, a language teacher, spoke eloquently about the use of traditional characters as a form of identity, contrasting the Taiwanese focus on the traditional characters and long history of Chinese culture, with the simplified characters and (she implied) the rejection of Imperial history on the mainland. However, in Toronto, since the dominant Cantonese community also preferred traditional characters to simplified ones, these did not serve to differentiate the Taiwanese community as a whole, so the use of traditional characters was more a way of expressing a diasporic or "overseas Chinese" identity than a particularly Taiwanese one. Finally, in Taipei, while most characters on signs were traditional, simplified characters were often used without exciting comment or concern (and there were visibly at least three systems of romanization in common use). While food was an important symbol in all three cities (see chapter 3), London emphasized Taiwanese food as a distinct culinary tradition, while Toronto practised fusion cuisine, and in Taipei the emphasis was on gourmet food of any origin.

The use of identity, expressed through symbols, can be clearly seen as the basis of the network society, for it can be expressed in ways that are common to all members but that also allow different meanings to emerge in different contexts. Everywhere, Taiwanese are united by the same symbols, but their meaning and importance depend heavily on context.

The identification with more extensive groups, such as the Chinese diaspora and those of other Asian countries, also deserves some unpacking. Seen from one perspective, this could be seen as a small, less-powerful group identifying with a larger and more recognizable group in order to gain visibility and develop networks (see, for instance, the Benelux nations' relationship with the European Union). From another, however, it could be seen as using shared points of identity to create transnational connections: appealing to a shared Chineseness, for instance, in order to develop business contacts on the mainland or elsewhere in the diaspora. It could also be seen as a way of combining different identities on different levels through self-presentation, as a means of negotiating the complexity of the migrant experience. However one views it, this combination of ethnic and, as it were, meta-ethnic, identities provides a different perspective on the use of identity in transnational contexts, pointing to the fact that sometimes a meta-ethnic or pan-ethnic identity may be employed as well as, or instead of, more traditional ethnic identities.

To sum up, the Taiwanese showed different types of transnational engagement in different cities: sojourners in London, settlers in Toronto, digital voyagers in Taipei. They also had clearly different experiences of migration, and different ways of using identity and self-presentation, and, while they all combined ethnic with other identities in order to build connections, they did so in different ways. The case of the Taiwanese shows the effect of time and place on migration and also points to the existence of different globalizations: it is not just that globalization has been different in different times, as with Foner (1997), but also that different globalizations exist, and are experienced, simultaneously in the present.

While studies of globalization have noted from the beginning that life in the network society inevitably entails navigating a combination of local and global spaces, the comparison of the three cohorts of Taiwanese migrants shows that the differences between locations have a crucial impact on the nature and experience of transnationalism, without making the people involved any less "transnational" in any particular place. All three of the sites of the study are "global" in different ways, and have different histories of globalization, and the experience of being Taiwanese is also different in each location. All of this indicates that life in a network society is not a single condition or experience, but rather entails a variety of globalized activities existing simultaneously. The question remains, however, whether the same is true of transnational social forms. To that end, the following chapters explore different kinds of globalizing organizations and the roles they play in linking the nodes in the space of flows.

Chapter Eight

The Social Network: Migrants and Transnational Networking Organizations

Having explored the roles played by location and by ethnic identity in the networking activities of Taiwanese skilled labour migrants in the UK and Canada, as well as for their counterparts in Taipei, I now consider a final aspect of their experience as transnational skilled labour migrants: participation in formal or informal networking organizations based, wholly or in part, on ethnic or national identity. As with global cities and the diaspora, the role of such organizations has long been discussed in the literature, but the differences between them, and the complexities of identity they reflect, are not usually examined in detail. The multiplicity of identities and networks makes for correspondingly multiple, interconnected globalizations, suggesting that new approaches to studying the network society are needed.

A Place to Stand, a Place to Grow: Networking and Transnational Social Organizations

The use of networking associations by migrants, labour and otherwise, has frequently been discussed in the literature, a useful summary of which can be found in Cordero-Guzmán (2005). Cordero-Guzmán states that such groups usually organize around ethnic or national identity (p. 893). They provide immigrants with the social and institutional infrastructure required for them to adapt or integrate to the host society, as well as formal assistance with transportation, housing, and legal procedures (pp. 902–3), and they can serve as liaisons with the country of origin (p. 907). He notes that "organisations help immigrants adapt while at the same time, often very deliberately, they help to maintain specific cultures and traditions" (p. 903). Schrover and Vermeulen (2005) note that host governments can also play a role: "Governments of the host society can use immigrant organisations to mould immigrants into a coherent community. This makes it possible to address the community, and hold it responsible for its members" (p. 826).

There has been less consideration, however, of transnational influences on these organizations. Moya (2005), for instance, dismisses the role of political groups, writing that these "are more likely to be local clubs, or at most national federations, than institutions like the (old) international Communist Party or the (new) American Republican Party, whose categorisation as voluntary associations is more problematic" (p. 835). Although he does allow that sending governments can be the sponsors of ethnic organizations, he does not consider the role that such tacit political power may play in influencing clubs. This does not entirely square with the numerous examples in the literature to the contrary, for instance, the connections between the Irish Republican movement and the Irish diaspora in the United States (Wilentz, 1979; Lynch, 2009), or, in the Taiwanese context, the KMT looking to the Chinese diaspora for support in the 1970s (Yu & Bairner, 2008). It therefore seems clear that immigrant organizations play a useful role in allowing integration while enabling migrants to maintain transnational links, and that these are also subject to complex, even competing, external influences.

Less has been written about organizations that are aimed at specific ethnic groups and that focus on migrants without being directly set up as "immigrant organizations," such as student and alumni organizations, or chambers of commerce. Although Moya (2005) includes discussions of groups such as hometown associations, religious organizations, and mutual aid societies, he does not consider related groups, such as trade associations, chambers of commerce, or alumni or student groups. One might add that traditional craft, sports, and performance organizations, such as martial arts clubs, choirs, and traditional dance societies, also tend to be associated with and/or dominated by specific ethnic groups. The discussion of migrant organizations thus tends not to consider the case of organizations that may combine one identity with another, or that may be focused on specific symbols of identity (such as food, dance, or music).

Furthermore, little attention has been paid to the potential differences between, and complex networks generated by, such groups. Schrover and Vermeulen (2005) consider that many factors may affect their formation – "the migration process ...; the opportunity structure in the host society; and the characteristics of the immigrant community (of which level of resources is just one element)" (p. 826) – and that they are affected by relations with the host society and its institutions (p. 831). Moya considers the debate over whether immigrant and immigrant-focused organizations are assimilationist or intended to preserve the pre-migratory culture (p. 837), when it seems quite possible that such organizations can be both at the same time, and indeed can be the (possibly unwitting) agents for the development of another, transnational culture. The complexities of such organizations, and the different globalizations generated by them, must be considered.

Finally, some consideration must be given to the online component of networking. Beginning with Miller and Slater's (2000) pioneering study of Caribbean migrant groups on Usenet, a growing amount of research has been done exploring how labour migrants use online activities as a complement to their offline networking. Hall (2003), in his study of the Usenet group soc.culture.taiwan, explored ways in which online networking has been key to the Taiwanese diaspora's efforts to define itself and maintain its identity since the early 1990s. More recently, Montgomery's (2008) study of Taiwanese skilled labour migrants, Ryan and Mulholland's (2014) of French migrants, and Marlowe's of the use of social media platforms by refugees settled in New Zealand (2019), have explored the use of networking and communications apps and sites such as Facebook and Skype as important tools for developing transnational connections, allowing organization members to communicate across distances and arrange offline meetings. Space does not permit an extensive discussion of the role of online groups in building and maintaining the network society, but since the Internet is undoubtedly a resource for a generally computer-literate and well-off group such as the Taiwanese skilled labour migrants considered here, some aspects must be discussed.

Like the ethnic networks discussed in the previous chapter, networking organizations create cross-border social connections that seem straightforward at first but are more complex on further inspection. As with the global cities considered in chapters 4 and 5, these organizations also contribute to the development of different globalizations through the different networking opportunities they present, using identity to bridge the global and the local. Networking organizations are connected to global cities and ethnic networks; simultaneously, they are part of the space of flows in their own right.

Making Money and Making Friends: The Role of Business Organizations

Connections within the Community

Business organizations, such as chambers of commerce, merchant organizations, and trade organizations, allowed networking and connections to be built through a combination of ethnic and professional identity.

Most of the members of the Taiwanese chambers of commerce whom I met generally had as their primary aim to promote their own personal business or advanced business services firms and were using shared ethnic origin as a means to develop business connections. The different business organizations all had different levels of official, and unofficial, political involvement: the chambers of commerce, for instance, were not government-sponsored, but local diplomatic organizations did maintain unofficial friendly contact with them, while the Taiwanese External Trade Development Council (TAITRA) was officially supported by the Taiwanese government; by contrast, the Taiwanese Merchants Association of Toronto appeared to be entirely locally driven. Different organizations, therefore, have greater or lesser degrees of political connection and influence and also vary in the degree to which their focus is local or transnational.

When asked about the role of such organizations, Michael, an official of the junior chapter of one of the Taiwanese chambers of commerce, cited some fairly prosaic reasons that apply to many more people than the Taiwanese:

> More and more young Taiwanese want to create their own business and to have their own business. When you start up the new business [it] will be quite difficult to get into the new market, so if you can leverage in this kind of social networking, probably for you to start up in a new company you're probably new to the content, you need an adviser, but in the junior chapter probably we have some specialized – they are quite professional, in such kind legal, in such kind for property, such kind for accountant, so we can leverage each other to coordinate it.

However, a later response in the same interview also indicated that the specifically Taiwanese affiliation of this group might produce a sense of mutual obligation that would make for a greater willingness to help one another, with a clear subtext of *guanxi* networking:

> INTERVIEWER: So it's partly about communication and partly about giving your own people a hand?
> MICHAEL: Yeah, I think just do them a favour.

Michael also, however, indicated in the material quoted in earlier chapters that Taiwanese people would understand the specific cultural and institutional hurdles that their co-ethnics might face in the UK: obtaining visas, dealing with bureaucracy, making the right social connections. It is not simply a case of finding a business partner or mentor – that could be done through any number of channels. The specifically Taiwanese nature of the organization meant it was possible to find people who would understand specifically what the person in question needed.

Outside Connections

However, the appeal of such organizations for interviewees was not simply their Taiwanese network connections; they could also be used to build networks beyond the Taiwanese community. Indeed, as Rupert indicates here, they could also be useful in circumventing some of the political problems caused by the businesses's Taiwanese origins:

> Because we [i.e., Taiwan] do not have diplomatic relations with a lot of the countries, so [the World Taiwanese Chamber of Commerce] is a major force and especially when we talk about Taiwanese companies, they're not only based in Taiwan but they go beyond Taiwan and in fact, around the world. And in many countries, they are a major business force in their respective countries.

The complicated relationship between business organizations and political infrastructure thus allows the former, through their mandate to further the country's business interests, to act as unofficial channels of political influence in contexts where the official political stance is not to recognize Taiwan as an independent country. Michael, similarly, indicated that while the organization's business connections went back to Taiwan, the aim for many members was to develop and maintain connections in the UK, with both Taiwanese and British partners, and thereby extend the bamboo network's borders to include networks beyond the ethnic group. Furthermore, in a city like London, where people are encouraged to follow the sojourner pattern and ultimately to return to Taiwan, the chamber of commerce could allow them to maintain their British business connections abroad.

Rupert's comments here are also worth considering in light of the differences of location and network discussed in the previous two chapters:

> For example, in the UK, you know, I assist in bringing major companies like BenQ, like Acer, all that, to set up their factories in the UK. And also, therefore, it is from the local MP's point of view, they create jobs in their constituency. And from the Taiwanese government point of view, the Taiwanese company do[es] not have to go to mainland China, like what they have been doing in the past because they speak the same language.

> ... And, for instance, when they come to UK, from the Taiwanese company point of view, I help them to get government grants from the British government, like [the] Regional Development Grant. So it's a win–win situation for the Taiwanese company, but it's also a win–win situation for the British government point of view, because it creates jobs.

Seen in light of the discussion of the impact of location in chapter 7, it is clear that the chambers of commerce are embedded in the regional context in which they are located and in particular discourses of globalization. Taiwanese companies in the UK must operate within the sojourner discourse – that is, they must set up operations in ways the British government can accommodate while also maintaining their networks back in Taiwan. But furthermore, in light of the discussion of social networking, Rupert, despite being a long-time KMT member and supporter of reunification, nonetheless indicates that he thinks it is a good thing for his organization's members to build ties away from the mainland – noting in passing that this is indeed where they tend historically to develop their networks – and instead build connections around the world. British discourses of migrant identity, and Taiwanese ambivalence about networking with the mainland, thus influenced the actions and orientations of the business organizations in the UK.

It is worth comparing the British and European organizations to the Canadian ones. Bruce, the representative of the Taiwanese Merchants' Association of Toronto, when asked what his organization does, gave responses that seemed at first very much in line with Michael's and Rupert's:

> It's a group of volunteer businesses; they, they all started in about, I think, twenty years ago, okay, and we were doing business here, but we need to kind of have clear paths for younger generations about what we do, and so we want to network with the business people in Toronto, and then also pass on the experience and also resources to the younger generation, and that's how the association or organizations been formed ...
>
> Companies, they can approach TMAT and say, "we are looking for such-and-such a person," and we would just say, "okay, I think there's a person I know, he owns a company in this field and he might be interested or he has people he refer you to," and so that would be for career opportunities for people or a lot of times that we do a lot of non-written contracts, okay, it might be good to start some business and the other person will say, "okay, I'm interested," and they say, "how much do you need?," "I need such-and-such an amount," it's totally verbal but it's because it's a tiny small community, we know each other so we pretty much take it as a contract and say,

> "okay, you just said you want in, how much, what's going to be your share, what are you going to do?," and so that's a, so the community type of closed relationship, it's very strong in Taiwanese business community.

However, although the organizations clearly have the same basic mandate, with the TMAT also helping its members develop connections in Asia and elsewhere as well as locally, there are some interesting differences in presentation and emphasis here. In keeping with the settler discourses outlined in chapter 5, for instance, the focus is clearly on the Taiwanese as a small ethnic "community" in Toronto, and the networking Bruce emphasizes is within the Toronto business community, in contrast to the sojourner-discourse approach, which is to work within local government restrictions and then maintain the resulting overseas networks after the return to Taiwan.

Furthermore, there is an echo of bamboo network discourse in Bruce's discussion of how the close-knit nature of the community allows for verbal contracts and trusting relationships. Elsewhere, he told me that "it's primarily in Asian communities that this kind of relationship, we call it, relationship is very important." There were clear connections with *guanxi* networking and ethnic networks in London as well: Michael's remarks aside, during the dinner at the European Taiwanese Chamber of Commerce meeting that I attended, candidates for official positions within the organizations went from table to table, offering drinks, proposing toasts, and generally rallying support and building potentially useful political networks. The organizations are affected not only by local identities but also by their members' construction of cross-border ethnic networks.

Hybrid and Complex Connections

It is worth unpacking the complex intertwining of the local and the global in these organizations. In the TMAT case, Bruce was speaking of developing and replicating those ethnic networks within the community, while Rupert spoke instead of extending networks beyond the Chinese and/or Taiwanese community, and the London cohort did not seem to emphasize the mutual trust engendered by the small size of the community as much as the Canadians did. This does not necessarily mean that relations were more formal; rather, they were discussed in such a way as to suggest that they were. The different globalizations of London and Toronto clearly affect social networking, the role of organizations, and the construction of ethnic networks, simultaneously and in different ways.

Furthermore, in both cities, Taiwanese business organizations could be used to build networking ties beyond the Taiwanese community. Rupert, discussing the World Taiwanese Chamber of Commerce's decision to donate money to flood relief in Haiti, said, "so they become a major force in helping not only Taiwanese people, but obviously countries who have got diplomatic relation with Taiwan, [countries which are] poor." TAITRA built connections with business organizations from other Asian countries in the London area, putting on joint trade fairs with Korean and Japanese trade organizations. This evoked the pan-Asian discourse of Taiwanese identity and its colonial connections to Japan. At the annual meeting of the European Taiwanese Chamber of Commerce, I encountered Malaysian Chinese and Swedish business people, who were attending the meeting in order to make business contacts among the Taiwanese business community (both in the UK and overseas). TMAT also collaborated with businesses from other Asian countries, with Bruce naming Korea as an example. He even indicated that the Koreans had an interest in "relationships" similar to that of Taiwanese and mainland Chinese business people, suggesting that they shared a *guanxi*-like networking approach. An ostensibly monoethnic networking organization thus was also used as a means to connect with people from other ethnic and national groups, through common identities and shared connections.

Finally, other identities combined with ethnic identities as part of the groups' operations. Michael, for instance, indicated that professional and age-related identities also affected the activities of members of his business networking organization:

> That is why we want to combine this kind of social networking, no matter [if it is] the Senior and the Junior Chapter [of the organization]. Senior [Chapter members], of course they have a very successful business, they will take out any business chance, they want to hire new people, either a new website designer or hire the new marketing, they can check with our Junior Chapter [where] someone is very proficient in this field.

The European Taiwanese Chamber of Commerce combined the European and Taiwanese aspects of its identity. For instance, the members sampled local Scottish cuisine during their annual meeting in Edinburgh at the same time as they enjoyed bubble tea, a specifically Taiwanese delicacy that two young entrepreneurs were then introducing to the UK. This practice links to the use of food – specifically, a cosmopolitan attitude toward cuisine – as a signifier of Taiwanese identity elsewhere. The ostensibly ethnic focus of these organizations thus belies their complicated links to other groups and identity discourses.

Taiwanese business networking organizations, therefore, not only allow members to construct transnational networks through shared ethnic identity, but also provide ways of extending those networks out in other directions, and of combining ethnic identities with professional, age-based, regional, and other identities. Furthermore, they affect, and are affected by, local discourses of globalization and the construction of transnational ethnic networks.

Old School Ties: Student and Alumni Organizations

University-Linked Organizations and Transnational Networking

Student and alumni organizations can be used to develop transnational networks through shared ethnic as well as institutional ties. This should not be surprising, given the role that international education has played in the development of the Toronto and London cohorts. As with the business organizations, university-linked organizations led to complex interconnections between local and global identities.

In the literature, Montgomery (2008) describes how Taiwanese migrants in the technology sector in the United States made use of their alumni connections, and the email networks associated with their former universities, to build transnational connections, find new jobs, and support globetrotting careers in ways that their American colleagues did not. Earlier, Li (1994), in the Chinese context, argued that the benefit of going to an elite university is as much about network connections as it is about actual education. But Li's study did not consider that such organizations, in the case of the Taiwanese, showed a pattern similar to that of the business organizations discussed earlier: they were nominally focused around ethnic identity, but in practice they also led out into other identities in complicated ways. Also, they could be affected by location and by ongoing transnational ethnic networks, and like the business organizations, they had differing political affiliations and levels of involvement. The literature indicates that student and alumni organizations showed a pattern of enabling transnational networking similar to that of the business organizations.

Among the communities I studied, student and alumni organizations operated, first and foremost, to support Taiwanese students in the host country as well as the alumni of their universities (Taiwanese and otherwise) worldwide. Although "networking" was not explicitly flagged as part of their function, this was clearly the case: representatives spoke about organizing social events and parties, with Daniel, the representative of the national-level Taiwanese Student Society in the UK, referring to the society as a "social platform." He alluded to the networking aspects of membership: "Say, after you finish your study you want work here or you want to develop, I don't know, other areas, so you probably need to know a few more people." The students may use the student societies mainly to make new friends and have a good time, but undoubtedly, they also use those societies to construct networks that are to their advantage either immediately (in terms of negotiating the complexities of the university experience) or in the future (in terms of meeting useful contacts for their subsequent careers).

Similarly, the alumni engaged mainly in social activities such as reunions and parties, and their primary, official focus was on allowing former students of the university to keep in touch with one another and to support the university itself financially. However, there was the obvious secondary benefit of building and maintaining key professional connections, both among one's former university connections and with people in previous and subsequent student cohorts. It is worth noting that all of my interviewees in Taipei, and quite a number in Canada, were contacted either directly or indirectly through student and/or alumni organizations. Clearly, such groups are crucial to the task of constructing social networks built around ethnic or ethnic-related identity.

As was the case with business associations, *guanxi* and ethnic networks affected university-related networking. Mrs Wang, explaining how she was planning to contact interviewees for my study, said of her fellow students, "They call me *jie jie*, older sister. I remember every single ... classmate, I know because we are in uni all the time." Her use of "older sister" evokes fictive kinship ties, and the idea of familial obligations within the network, and when considering who else to ask to participate in my study, she referred to "doing favours" in a way suggesting a *guanxi*, or *guanxi*-like, system of social obligations. There was a clear transnational element as well: a significant number of the people I met and worked with in Taipei had been to university in the United States, the UK, or (in a few cases) elsewhere in Europe or North America, and this was generally considered to be a normal, career-minded, part of professional development (see also Teo, 2011, p. 812). Furthermore, I was told by a representative of a business organization in the UK that the Taiwanese design industry regards a degree from one of the top British art colleges, such as St Martin's School of Art and Design, as extremely helpful for career success. Student and alumni activities allow the building of transnational networks and in many cases the establishment of transnational careers.

The Complex Ties of University-Linked Networks

However, the Taiwanese student societies were not *exclusively* by and for Taiwanese students. This was aptly summed up by the head of the Taiwanese Student Society in London, who, when asked to explain what his organization did, said:

> We aim to, first to serve students, mostly Taiwanese students of course, *we also help other students who are also ... who are interested in Taiwanese related issues,* and secondly we promote Taiwanese cultures in various ways, sort of we support and what we call collaborate with Taiwanese artists who want to have an exhibition or a show and have some of their products. *As well as promote food cultures, as you are aware of Taiwanese food fairs, to European or other people,* yeah. (Daniel; italics mine)

The student society representatives, and recent students, with whom I spoke all noted that Chinese students from elsewhere in the diaspora, or from the mainland, would often attend their events, and vice versa; as noted elsewhere, many of the attendees at the Taiwanese Food Festival were not Taiwanese. The social aspect of such societies in general was clearly not restricted to Taiwanese people. As with business organizations, then, membership in the student organizations clearly allowed networking with members of other groups. One might notice, again, the overlap with the construction of *guanxi* networks, given the social aspects of student society activities and the likelihood of Chinese students from elsewhere attending.

Furthermore, some of the Taiwanese student societies had ties with government organizations, through the sponsorship and support of the Taipei Representative Office, but also with the alumni associations of the various universities with which they were associated:

> And from alumni, they run really good in Taiwan and usually they provide scholarships, although it's not much, but it's really helpful, like there is everyone got £100 a year, and there's a £4,000 award ... And I represent them ... Last summer I went back to Taiwan to see them and we had a meeting ... Report, [I] submit like a report to them, [to explain] what's going on really. (James)

In Canada, although I could not determine whether they had a similar level of direct government support, National Taiwan University (NTU) alumni organization officials could use their affiliation to develop significant political contacts. One showed me pictures of himself with top politicians from both Canada and Taiwan, including a Canadian ex-prime minister, as well as pictures from sports events and from visits to the Taiwanese Economic and Cultural Office in Canada, and said that these contacts had come about through his role in the organization. On a less political level, networking could happen in unexpected ways through university connections. In Taipei, one of my research assistants surprised me by revealing that she knew Maria, the travel agent in London profiled in chapter 4, because NCCU made frequent use of Maria's travel business to arrange academic travel and visits from prospective students from the UK. The research assistant told me that she favoured this business in particular over any other British travel agent because "they are Taiwanese. And they speak Taiwanese." Given the way in which language frequently crops up as a symbol of Taiwanese identity, to "speak Taiwanese" here seems to be a metaphor for understanding, or being sympathetic to, Taiwanese culture. The academic organizations allowed the construction of socially and/or politically advantageous networks both within and beyond the Taiwanese community, and across national borders, through formal and informal connections at various levels.

However, the Taiwanese aspect of the student societies could be complicated by other identities, particularly that of the related institution, as suggested by Li (1994)'s study of elite universities discussed earlier. While NTU has a clear connection with Taiwan, overseas universities such as those in Britain and Canada create transnational connections straight away among their Taiwanese alumni. Furthermore, this could be complicated by the distinct identity of the institution itself, as my interviewees at the University of Oxford indicated: Oxford maintains a strong presence around the world as an international elite university with a socially active alumni community in most countries, and the Taiwanese branches of the Oxford student organization must be taken in that context. To clarify, Oxonian alumni organizations are as much about the transnational identity of Oxford itself and of former Oxonians, as they are about the transnational connections of the Taiwanese. Taiwanese alumni of British universities, several of whom I was able to speak with in Taipei, maintain a transnational identity even after they have returned from their studies overseas.

The differences in local context discussed in chapter 7 also affected student and alumni associations. For instance, asked about whether the Taiwanese alumni of British universities in the UK were supportive, Daniel replied: "Yeah, we do have a few, but as you're aware, like, because the difficulty to issue work permits, to get permanent residency here, so our ... that alumni in UK is ... I'd say it's even less, I mean in terms of numbers than in Taiwan 'cause obviously we can go back to Taiwan any time we want."

The British focus on a sojourner model of migrant identity clearly affected the sorts of contacts that Taiwanese students in the UK had with alumni of their institutions. The Canadian group, in line with the more settler-focused discourses in their host country, spoke less about negotiating the migration bureaucracy or establishing a business, and more in terms of maintaining cross-border relationships with their co-ethnics in the United States. Again, host-country context affects the nature and focus of student and alumni networking.

Note also that there was no single pattern of involvement or networking within these societies or within their actual and potential memberships. Daniel reported that an attempt to provide a joint networking event between alumni and students earlier that year had been poorly attended but that his counterparts at Oxford University maintained active and lively ties with their alumni association. Not all students, and especially not all alumni, played an active role in their organization.

Age could also make a difference in terms of the type and kind of networking that participants engaged in. During my fieldwork, I found that alumni associations were much more fruitful sources of contacts than student associations, clearly suggesting that the young adults were less interested than older people in actively building long-term networks; instead, they used the alumni network to maintain social ties. So it is worth remembering that organizations are not uniform, directed entities, but are subject to change over time and to differences in the ways in which members perceive and use them.

The student and alumni associations, like the business associations, are, therefore, key sources of networking related to members' ethnic identities. Like the business associations, furthermore, they are affected by local discourses of migration and by transnational ethnic networks. Also, the student and alumni associations were used as much to construct ties outside of the Taiwanese and wider Chinese communities as within these, and their role as ethnically focused networks was complicated by the introduction of identities as students and graduates of particular institutions. It remains to be seen whether the patterns observed offline also hold true for online networks and social groups.

In Your Facebook: Online Networking and Identity

Social media also played a key role in the construction of personal and social networks. My informants used Facebook to maintain ties with relatives, friends, former classmates, and former workmates who now lived abroad or in different countries; it was also used by Taiwanese businesses to advertise their products and by various social and student groups to announce or promote activities, as well as to share news articles of particular interest to students and/or expatriates. The now defunct online forum Dim Sum, for Chinese and ethnically Chinese people living in the UK, while less prominent as a community organization at the time of my fieldwork than it had been in the past, was still being used at the time for seeking advice, posting events, or discussing issues; it was recommended to me by an interviewee in her twenties. The Taiwanese would supplement and reinforce their offline social activities with online ones, and vice versa.

As with their offline networking, however, my informants' online networking also led to the development of connections beyond the ethnic group. If anything, this was even more possible online, for the ability to join different groups and participate on different sites is inherent in the medium. A typical Facebook user might simultaneously be a member of a Taiwanese student organization, a non-Taiwanese student organization, professional organizations, organizations that might or might not have ethnic aspects (such as sports fan clubs, book groups, or cinema clubs), and organizations that involve no ethnic aspect at all. The flexible and multivalent properties of identity allow people to combine and express different identities at the same time in the offline world; the same can be said of the online world.

The use of Facebook and similar social networking sites is, therefore, particularly interesting from an identity perspective. Such sites are identity-based at their core: it is easy to join or create groups based on combinations of identity, such as Taiwanese/student, or Taiwanese/mother, and also for a user to build an eclectic image of his or her own identity by joining different groups. Facebook allows users to post both an English and a Chinese name if desired, allowing people to express Taiwanese or ethnically Chinese identity symbolically in an English-language context. Such sites encourage people to build connections with mutual friends, thus enabling them to expand their transnational social networks. Younger informants in particular were heavily involved in organizing through Facebook; some of them were also involved with KILTR, a now-defunct social media site focused on Scotland and, significantly, both the Scottish diaspora outside Scotland and ethnic minority diasporic groups (such as the ethnic Chinese) within Scotland. People who have grown up using the internet to express their identity find that it serves as a fairly natural extension of one's identity development in other ways. The fact that such sites can be accessed in most places around the world means they are particularly useful ways of building and maintaining identity and networks across borders.

Social media also enable a blending of personal and professional identities. Most of my Taipei interviewees were involved, in one way or another, with the IT and social media industries, as was one of my Toronto interviewees, and the fact that internet sites are used interchangeably for personal and professional identity expression (joining a Taiwanese expatriate group in order both to express one's identity and to promote one's small business, for instance) means that identities can be combined and extended across borders in complex ways.

For all of these reasons, such sites are useful for cross-border networking, along the lines of the migrant associations and student groups discussed earlier. While in Taipei, I used my Facebook account to post about my experiences and to gather knowledge from friends who were either Taiwanese or had been expatriates or students in Taiwan, who helped clarify my own misunderstandings and answer my questions about social practices. From such exchanges, I was also able to learn about things as diverse as how to navigate the Taiwanese satellite television system, the history of Taoist rituals, the local art of puppetry, and the use of different systems of romanization on the street signs, public notices, and so forth in Taipei. I also gathered more basic advice on which areas, restaurants, and cultural institutions to seek out and which to avoid. Internet forums and networking sites can thus serve the same function as more formal immigrant organizations, offering advice and introductions to migrants and allowing them to navigate the complexities of the host society.

The paradox is, of course, that even such transnational spaces as the internet are themselves tied in with location, so inevitably, host-country and other outside factors affect expressions of identity online as much as they do offline. For instance, at the time of writing, some social media sites have been blocked in mainland China. Significantly, there was at one point an active effort among diaspora Chinese to allow their mainland family and friends wider internet access through enabling them to bypass government firewalls (BBC, 2004). Also, mobile telephony and broadband have uneven penetration as a consequence of government regulation and cultural factors. Sites and networks can be developed that reflect quite specific local interests, as with KILTR (see above; although the site did have overseas connections through the Scottish diaspora), but their longevity is impacted by external factors; over the course of my fieldwork and later on, many independent forums gave up ground to Facebook groups and Twitter accounts, for instance. In much the same way that online activities reflect offline ones, they are affected by host-country and outside ethnic networking in complicated ways.

The internet, again, allows the development of an ethnic identity into a global identity, through the use of media that are visible and accessible in a number of countries while catering to specific local interests. Ethnicity can thus transcend national boundaries and become global even while being influenced by local factors and other forms of networking and identity.

Only Connect: Online and Offline Networking Organizations in Perspective

Networking organizations of various kinds, online and offline, strongly affect how transnational ethnic networks are constructed and maintained. The organizations I examined for this study were specifically identified as Taiwanese and focused on a particular ethnic identity, but, at the same time, they were helping their members form networks based on their roles as business people, entrepreneurs, or students. Moreover, these organizations combined professional and personal identities of various kinds in order to achieve this aim; online networking in particular allowed various expressions of identity to combine. Networking organizations, to fulfil their functions, build not only on particular national and ethnic identities but also on the flexible expression of these identities.

Furthermore, networking organizations link two prominent themes of this book: ethnic networks, and the role of location in transnational migration. While it was less easy to directly compare organizations between the cohorts, as different organizations tended to dominate my sample in different locations, there were clear similarities and differences between the use of business-focused and academic-focused groups in the British and Canadian cohorts. The form, activities, and mandates of the different groups discussed are, however, all affected by location-specific social and political discourses, for instance whether a "sojourner" or "settler" pattern of migration is viewed as "normal." Thus, while academic societies in London and Toronto generally focused on building and maintaining social connections, the ones in the UK were generally premised on a return to Taiwan after completing one's studies rather than on maintaining local networks, and TMAT in Toronto was more focused on connecting businesses in the Canadian context than on navigating a complex and hostile bureaucratic environment such as the London business associations. Location affects the activities of networking organizations as much as it does the expression of transnational identity in general.

Business and student associations are also linked with transnational ethnic networks. There were elements of *guanxi* networking in the way Mrs Wang spoke about her alumni organization and in the way in which the chambers of commerce and TMAT were used by their members to construct social networks with a view to supporting their business activities, allowing them to do so across national borders if needed. A number of the tropes of identity that emerged in earlier chapters were visible in the ways people spoke of, and used, networking organizations: the construction of Taiwanese identity as "ethnically Chinese" or as "pan-Asian" depending on the circumstances, the ambivalence toward China, issues of how to do business in a new society. Networking organizations combine discourses of localization and of developing global ethnic networks and also provide means for expressing discourses of Taiwanese identity.

It should be emphasized once again that the role of the different organizations was complex, as well as shot through with other forms of identity. Despite their names, none of the organizations I was able to make contact with dealt exclusively with Taiwanese identity, and only one with Chinese identity; all of them involved at least one other identity, as business people or as students/alumni of particular universities. In the case of some other organzsations, such as Buddhist or Christian organizations, the group crossed ethnic lines to work with and include co-religionists of different ethnic groups; as noted elsewhere, they also could be unofficially age-focused. As seen in the previous two chapters, identity allows people to combine different identities in complicated ways that extend into, and out of, the ethnic community.

Finally, it is worth noting that this is likely the case with the more directly targeted immigrant organizations discussed in earlier literature (Cordero-Guzmán, 2005, Schrover & Vermeulen, 2005). The organizations considered by Cordero-Guzmán and others are, arguably, tacitly combining identities – people of a particular ethnicity, in a particular city, and self-identifying as migrants (since not everybody who migrates does) – and are undoubtedly affected by local discourses about migration, and linking with diasporic networks and the ongoing discourses, referenced earlier, of what constitutes a diaspora (Ho, 2009). The Chinese community organization I was able to investigate as part of this study was open to Chinese from all over the diaspora, and included non-Chinese in such activities as Mandarin classes. Even when this is not explicit, therefore, organizations combine ethnic identity with other identities.

This examination of Taiwanese skilled labour migrants thus supports the general premise that being part of network society is a complex, multiple state of being affected by location, personal characteristics, and historical context. The different ways in which the different organizations engaged with the global and the local, and in which their members took advantage of them to do so, reflects the pattern of different globalizations; it also shows links between local and global discourses. As elsewhere, we have here seen complex, individualized globalizations in which the strategic use of ethnic and other identities to form local and transnational networks leads to a variety of patterns of globalization and localization within the same skilled labour migration diaspora.

An examination of the use of organizations, online and offline, shows again the pattern of multiple globalizations previously seen in the relationship of the Taiwanese to the global cities in which they live, and in the use of ethnic networks. A number of factors make a clear difference to how the network society is experienced: individuals' opportunities, their age cohort, and how they relate, demographically, to key world events such as the end of the Cold War and the start of the War on Terror (or more locally focused ones such as, in the Taiwanese case, the 1949 migration of the KMT from the mainland or the end of the dictatorship), the nature and kind of their global and local connections, relationships to other transnational migrants, profession and academic status, and a variety of other characteristics. As with the differences between global cities, this is frequently acknowledged in the literature, but the implications are seldom explored in great detail.

The Taiwanese skilled labour migrants, seen through three different lenses, provide us with a snapshot of a group that is fully integrated in Castells's network society, maintaining connections to different locations and navigating the social and physical spaces between them. Significantly, what we see is a single group that is, nonetheless, strongly differentiated by factors such as location, age, profession, and ancestry, indicating that a complex approach to studying labour migrants and other transnational groups is needed. Furthermore, the controversies of Taiwanese identity, and how it should be defined, are discourses that cross borders and both affect and are affected by the migration experience. All of this will be explored in greater detail in the concluding chapter.

Chapter Nine

Taiwan in the Net: Identities in Perspective

To bring the story around full circle, I now take a final look at the issues that drove the study of Taiwanese skilled labour migrants. First, I consider the complex nature of Taiwanese identity, not as a kind of ideal or substitute Chineseness, but as an identity in its own right, one that incorporates Chinese, pan-Asian, and local symbols of belonging. Then I return to the concept of the network society and consider what it looks like two decades into the twenty-first century, when challenges to globalization on one level are further embedding people in the space of flows on other levels. Finally, I formulate the concept of globalization as a differentiated social activity: as neither a unified placeless phenomenon nor a mass of global and local discourses informing one another, but rather as a condition that varies in different times and places, taking different forms that interact with and change one another.

The preceding chapters have examined the uses of identity by two groups of Taiwanese skilled labour migrants and the infrastructures that support those uses. This involved examining the complex and contested discourses surrounding Taiwanese identity, in particular how it develops connections between transnational and Chinese identities through ideas such as *Zhonghua* and the debate over whether Taiwan is Chinese or Southeast Asian. I then developed the concept of the network society as a research tool for considering transnational groups as nodes in a network, one in which various concepts of being local and global are experienced and reinterpreted by skilled labour migrants. The social infrastructure through which this activity persists and develops, such as business support groups and online communities, enables skilled labour migrants to operate across borders even when legal and social barriers attempt to block them from doing so, a situation deserving of more scrutiny in the context of rising protectionism.

This study has sought to conceptualize the experiences of individual skilled labour migrants as nodes in a global network and to explore what these experiences can tell us about the social factors involved in migration, and especially skilled labour migration, in the general spirit of structuralist anthropology (see Holý [ed.], 1987), and not to present the subjects of the study as somehow representative of all skilled labour migrants (or even of all Taiwanese skilled labour migrants). In this final chapter, I investigate what transnational social formations look like from such a perspective: they reflect a complicated combination of location-based and network-based influences, situated in particular times and places. From the cases discussed here, other migrants, scholars of migration, and social and political bodies aimed at dealing with migration – in whatever capacity – may gain greater insights into the Taiwanese experience specifically and what it can tell us more generally about life in a twenty-first-century network society.

"Made With Pride in Taiwan": Taiwanese Global Identity with, and without, China

At the start of this book, before situating modern Taiwan in its historical and political context, I said that the study would, in part, explore and describe members of an Asian diaspora that is widespread and influential and that has a complex and contested identity both at home and abroad, but that is understudied relative to its more powerful neighbour, frequently being either conflated with China or interpreted as a kind of pure, communism-free, Chinese identity. I will now consider what the study of Taiwanese skilled labour migrants can tell us about Taiwanese identities in the transnational context and their relationship to Chinese identities, particularly in light of the idea that Chinese identity, like many other ethnic identities, can be considered a global identity rather than one tied to a particular geographic region.

Taiwanese Identity: Chinese and Asian

Although officially the discourse of Taiwanese identity in the political sphere revolves around whether Taiwan is "part of (mainland) China," "real/pure China," or "Asian," this in fact engages with the concept of ethnic identity as a global social formation. If Taiwan is "Chinese," based on it having formerly been a province of China and on the majority of its inhabitants being of Chinese ancestry, the political assertions of the KMT, and/or the ethnicity of its dominant group (or, more accurately, groups), then the question remains – what it is to be "Chinese" (as

Chineseness would therefore have to include such traits as capitalism, republicanism, and an engagement with the imperial past)? However, if Taiwan is "Asian," then it is an Asian country that was once a Chinese colony, is associated with a breakaway Chinese political movement, and has a large ethnically Chinese presence. The question then becomes whether the Chinese in Taiwan are simply the largest ethnic group in a multicultural, postcolonial country that looks to Japan, Europe, and Korea for its cultural influences as much as to China, or whether Taiwan is a Chinese province with its own distinctive engagement with European and Asian colonial history.

However, this question cannot be answered, at least at present. This is not necessarily a bad thing. The contested nature of Taiwanese identity means it exists in a state of balance, or debate, or flux. For Taiwanese, the advantage of this is that it has enabled them to connect in a variety of different directions – to be Chinese, or Asian, or something else entirely, as the situation demands. While not always the most comforting or helpful situation – as witness Mrs Wong's heartfelt discussion of the confusion she felt about her own identity and her concern that it not affect her children – this can also be a strength in terms of forging global connections and adapting to various local discourses.

The Taiwanese diaspora faces a particularly complicated position, for most of its members are simultaneously part of the Taiwanese and the Chinese diasporas. In some contexts, Taiwanese identity can also be a political identity, as, for instance, with Rupert's adoption of Taiwanese identity as a means of asserting his ethnicity as Chinese together with his political stance as a KMT member. His active role in the community legitimized this aspect of Taiwaneseness, for he was able to promote activities related to the image of Taiwan as "a free China." Taiwan's position as simultaneously Chinese and not-Chinese meant that complicated discourses of Chineseness and not-Chineseness emerged, which were subject to debate, conflict, and attempts at control, but could also, in the diaspora, lead to the development of strong internal and external networks.

The different waves of migration meant it sometimes seemed as if there were multiple diasporas, with the Cold War–era migrants and the millennial migrants having significantly different experiences and attitudes. The older cohorts went abroad to benefit from foreign student programs and/or to seek a better economic future, or due to their stance vis-à-vis the KMT (some fleeing it, some going abroad to support its external activities); the younger cohorts generally went abroad to study, to work as expatriates, or just to see the world. Some reasons were held in common – such as to acquire a higher education or to live with a non-Taiwanese spouse or partner – but even these often took different forms depending on the historical context.

Relationships between Taiwan and its diaspora also varied with the historical context, as in the case described by Yu and Bairner (2008). Regarding London and Toronto, the composition of each city's respective Chinatown reflected different waves of migration as well as different relationships between the mainland and Taiwan, as noted in Mr Lee's observation about the London Chinatown (which also, in its phrasing, contains a tacit assertion of Taiwan's claim to Chineseness over that of the mainland): "Before that [i.e. in the twentieth century] I think some of the early immigrants, they come to here and stay and have their Chinatown, it's a long story so ... but at that time China is part of Taiwan, we still have the dominant power, but later on the China is getting stronger and stronger, then they shift." The situation in Toronto, where there were tensions between the Cantonese-dominated downtown Chinatown and other neighbourhoods dominated by Chinese migrants from elsewhere, reflected the importance of historical context for understanding ethnic identity. Historical factors in Taiwan, and in the global cities in which the study took place, thus shaped and influenced identity and migration.

In addition, the identity discourses of the host country clearly influenced those of the migrant groups. In chapters 4 and 5, I considered how London's identity as a "city of sojourners" and Toronto's as a "city of settlers" affected the ways in which Taiwanese people related to their surroundings, regardless of whether they themselves personally fit this paradigm. These discourses are rooted in wider concepts of identity. The UK, for instance, has strongly defined itself in recent years as *not* a nation of migrants: British politicians today frequently take an anti-migration line, which clearly affects not only official policy but also attitudes toward migration. This anti-migrant position, however, is in conflict with London's distinctive identity within the UK as a cosmopolitan enclave with global connections and active migration (McDowell, 1997). The "sojourner" discourse has emerged as a result, with migration seen as acceptable provided that it is temporary and focused on a particular goal (making money, acquiring higher education, etc.). This is further supported by the wider discourse that London is a difficult place to live even for native-born British, who are encouraged to seek quieter, cheaper locations in later life. While two of the British interviewees were in fact established in the north of England and the Midlands, this also reflects the image of these areas as traditional areas of manufacturing, particularly for Asian companies seeking to launch British operations (as pee Elger and Smith's [1994] studies of the area). Edinburgh, another key node in the Chinese–British network, is likewise global in its own right, if in a different way than London. Once again, the image of the migrant in the UK as a temporary expatriate reflects wider discourses about national identity and migration.

Toronto's "settler" discourse, likewise, reflects the discourse of the "Canadian mosaic" (Kelley & Trebilcock, 1998), that is, the idea that Canadian identity coexists comfortably with other, diasporic identities. A less positive aspect of Canadian identity that emerged was that of Canada's insignificance; for my interviewees, Canada was not generally a first-choice destination, but a place where people wound up staying on their way to the United States or Europe (even Nora, who had migrated to Toronto by choice, had initially viewed the United States or Europe as preferable destinations, before finding out about Toronto's assets through chance). Decisions about migration are thus affected not just by national discourses of identity, but by wider discourses about the nation in which they settle.

Taipei itself, meanwhile, constructed itself as a global city in terms of its diasporic connections elsewhere, and also in terms of the outward-looking aspects of Taiwanese identity; people spoke repeatedly of how Taiwan had always been a nation of seafarers and pirates. This is particularly interesting, for this is a Chinese take on Taiwanese identity, one that ignores the narratives of the island's indigenous peoples – who were not pirates and seafarers. Arguably, it also ignores less romanticized waves of colonization by the Chinese, Spanish, and Dutch in the seventeenth century in favour of one based on the earliest, sporadic, Chinese uses of the island (Andrade, 2007). It also, interestingly, constructs Taiwanese globalization in a way that nods toward *benshengren* more than *waishengren* identity markers. Taipei's architecture tends to acknowledge Japanese and European influences. Also, as Liu and Hung (2002) note, the teaching of Taiwanese history has been changing since the 1980s to acknowledge the non-Chinese influences on the country. Taipei's construction as a source of dialogue about what it is to be Chinese and Taiwanese, and the linking of both with globalizing discourses, also affects how people express their identities, domestically and abroad.

Other identities came into play in addition to the ethnic and geographical ones. University-related identity was significant for many of my interviewees, though not all of them. Attending Oxford, for instance, might be framed as a means of embracing a British identity; or it might be a global one, given Iyer's (2000) markers of a detached global cosmopolitanism. Religious affiliation, another potential marker of a local or global identity, also emerged in complex ways. Buddhism, for instance, was an identity marker – albeit more for older than for younger Taiwanese in London – besides providing connections with other Buddhists.

My evangelical Christian informants associated their Christianity with a transnational faith community and the idea of a global religion rather than with particular local identities. Ethnic, transnational, professional, and academic identities thus combine with Taiwanese–Chinese tensions to form complex patterns with different affiliations with the local and the global.

The complexity of the situation has effected the migration of skilled workers and their connections with other places. For instance, the received wisdom is often that Taiwanese businesses have strong connections with China (e.g., Deng, Huang, Carraher, & Duan, 2009; Tien & Luan, 2015), given that the Taiwanese are ethnically Chinese and, in the case of the *waishengren*, may have recent mainland connections. But in practice, many Taiwanese are ambivalent about the mainland, because they are *benshengren* or for other reasons; for instance, despite being *waishengren* with active family ties to the mainland, Glinda was a supporter of Taiwanese independence, and Sun's (2014) study found that *waishengren* returnees to the mainland were ambivalent about their experience.

Furthermore, this sort of complexity is far from unique or unusual. Many postcolonial countries, particularly those that have been multiply colonized, look to diverse colonial referents while at the same time struggling to assert their own distinctive identities (see, for instance, Herbert, 2012). There are similarities here with countries such as Canada and New Zealand, which are overshadowed by a more powerful neighbour yet maintain distinctive national identities even while sharing many cultural similarities and reference points. Even in more straightforward situations, there are echoes of Selmer's (2002) study of American expatriates in China, which found that ethnically Chinese expatriates encountered more problems, despite conventional wisdom suggesting that the opposite would be true, arising from the fact that they were assumed to be culturally "Chinese" when in fact they could equally hold American values and cultural norms. Identity, particularly in skilled labour migrant contexts, is a complex factor that may not always play out as expected.

One area that the literature on identity and migration does not sufficiently embrace is historical context and the influence of the past on the present. For instance, it is often acknowledged in the literature on globalization that there have been a number of waves of globalization, for instance, in the pre–First World War era (Foner, 1997). Less often discussed is that these earlier globalization events continue to influence globalization today; the present system has been influenced as strongly by nineteenth-century colonial globalization as by the end of the Cold War in the 1990s. Similarly, the Taiwanese diaspora in London and Toronto has been influenced not just by recent labour migration but also by migration events dating back to the nineteenth century; Taiwan itself owes some of its identity to colonial expansion events in the seventeenth and nineteenth centuries (see Andrade, 2007; Liu & Hung, 2002). The literature on globalization needs to pay greater attention not just to the immediate situation of migrants but to the wider background and context that led to that situation.

Zhonghua as Global Identity

Looking at different communities of Taiwanese around the world also exposes the complicated discourses of Taiwanese versus Chinese identity. In a diasporic context, the idea of a global Chineseness, a *Zhonghua* not necessarily related to *Zhongguo*, comes much more to the fore: my informants built relationships with fellow migrants from elsewhere in the Chinese diaspora, or sought business partners with an interest in building mainland connections through a Taiwanese entry point. This was further legitimized through the above-mentioned discourses of globalization in Taipei, which emphasized study abroad and mainland connections. One might also note that a sense of global Chineseness does not necessarily mean that one focuses on what all Chinese have in common: Hall's (2003) informants found that visiting the mainland confirmed their feelings that Taiwan was not "part of" China and had its own distinctive identity (p. 161). The connections between Taiwanese and Chinese identities were, therefore, complex, and part of an ongoing and always shifting discourse of belonging.

A few possible models emerged from this study regarding Taiwan's relations with its neighbours. There was, for instance, one in which Taiwan was equivalent to China (in an ironic twist thanks to the nature of symbols, this point of view was simultaneously that of the KMT and the mainland Chinese); and another in which Taiwan was a Southeast Asian nation along the lines of Malaysia and Singapore, with a large Chinese population but not in and of itself "Chinese." There is also Chen's (2012, p. 846) observation that many recognize a distinct Taiwanese identity without seeking distinct Taiwanese nationhood. National identity thus exists in transnational settings (Helbling & Teney, 2015) but does not always involve a relationship to a particular country; indeed, it may be subject to contested discourses over what form that identity takes.

All of this is far from unique to the Taiwanese or even the Chinese more generally. The *Zhonghua/Zhongguo* distinction brings to mind eighteenth-century pre-national concepts of Germanness (Watson, 1995, chs. 1–3). In Taiwan's relationship to the mainland, there are also echoes of the British Commonwealth, with countries dominated by an ethnically

British colonizing population developing distinctive identities over the decades and centuries even as they maintain cultural, social, and economic ties to the colonial country. The case of Taiwanese identity in transnational contexts may indicate wider traits of transnational identity in other populations, which are seldom considered, suggesting that many groups participate in identity discourses in a similar way.

As such, then, a consideration of the complexity, and management, of Taiwanese and Chinese identities among Taiwanese skilled labour migrants provides insights not only into the complexities of Chinese identity but also into the influences of local identities, and other identities, on diasporic identities, and into the need to situate migration not just in its immediate socio-historic context but additionally in terms of wider spans of time and place. Taiwanese identity should be viewed not as a "free Chinese" identity, or as a stand-in for Chinese identity more generally, but as a distinct ethnic identity that has associations with Chinese identities, Asian identities (both, themselves, complex and subject to a variety of internal and external discourses) and with global identities.

Acting Globally, Acting Locally: The Continued Rise of the Network Society

Studying Taiwanese skilled labour migrants also allows us to reconsider and further develop the theories and models put forward in Castells's Information Age trilogy. To begin with, it provides an in-depth practical example of a single networked group, at the nodes of the network and navigating between them. More than this, however, it clearly shows the role that identity plays in the network society, adding to Castells's examples by showing how identities change and adapt as part of complex discourses and how this is, in fact, the reason why identity is of such crucial importance in globalizing social spaces.

Identities and the Influence of the Local

The first point to consider is the multisited nature of the study, which, as noted, reflects a somewhat rare and new approach to anthropology. Studying the same diaspora in different locations means that one can consider how people from similar ethnic and socio-economic backgrounds fare differently in different social environments.

Both the destination cities studied here are "global cities" in English-speaking, OECD countries, with large populations of both elite and non-elite migrants. Both recognize Taiwan unofficially (or semi-officially) while officially acknowledging mainland China's claim to the island, and both recognized Taiwan as a political entity prior to the rapprochement between the US and mainland governments. Both cities have diverse ethnic Chinese communities that are, tacitly or explicitly, dominated by Cantonese.

The Taiwanese cohorts studied here embrace similar discourses of ethnic identity: they use many of the same symbols, such as food, traditional Chinese characters, Mandarin, and the use of Wade-Giles romanization (see Wang, 2004), and they both celebrate holidays such as the Mid-Autumn Festival and the Spring Festival (Chinese New Year). The interviewees in both locations raised many of the same discourses of identity, such as the history of *waishengren/benshengren* tensions and the question of whether Taiwan is more "Chinese" or more "Asian." Differences within cohorts were similar in the two cities, with older informants still speaking in terms of Cold War–era alliances and the "communist China/free China" debate, and younger ones focusing more on transnational identities and Taiwan's relationship to the wider Chinese diaspora.

Yet significant differences exist between the two locations, found, for instance, in local discourses about what constitutes appropriate behaviour and social engagement for migrants: London expects them to move on after a few years, while Toronto expects them to settle and bring their family members over (again, regardless of what was actually the case for individuals). London has a social cachet as a global destination that Toronto does not – hence my informants' desire to live in London, if only for a short time. By contrast, migration to Toronto tends to be accidental and not the migrant's first choice. London is viewed as a prestigious city where one can build a career, but also as a temporary stop; whereas Toronto is seen as a relatively safe place where one can raise a family. In both cities, Cantonese dominance is more of an issue – occasionally it verges on a rivalry – albeit more for the Torontonians than for the Londoners, who never explicitly mentioned it. Finally, there are the tacit influences of the home country and of Taipei's own construction of globalization. In Taipei, globalization manifests itself through the connections with the mainland at various times and places and through its citizens having worked or studied abroad, with the paradoxical result that connections with such cities as London and Toronto become part of national identity at home. A variety of local influences affect discourses of identity, combining and intersecting in unpredictable ways.

Since identity is a discourse between migrants and host populations (see Berry, 2005), it is also worth considering which identities are "allowed" in each context by the host societies. For instance, although Canada and Scotland both have skilled labour migrant populations – with the population of Canadians living and working abroad being larger than that of many Canadian provinces (CBC, 2009) – people do not normally think about these as migrant diasporas in the way they do about the Irish or Chinese diasporas. Contrariwise, Ho (2011a) points to groups that would not necessarily consider themselves diasporas being labelled as such by the home-country and/or host-country governments for political or other purposes. In this study, some interviewees were willing to accept outsiders' identification as "Chinese" rather than "Taiwanese," due to the complicated nature of diasporic identification, while others tacitly rejected the concept of diasporic connections; Kay, for instance, said he would not necessarily encourage his children to learn Mandarin despite their heritage. There were two cases in the British sample of people who were groomed to inherit the family business (one male, and the other, less traditionally, female) and who rejected this after having lived and worked abroad, despite their elders' conceptions of what it means to be Taiwanese or Chinese. Outside discourses of identity thus affect who is allowed to identify as a migrant, as well as what identifying in this way might involve.

Crucially, the cohorts also showed differences in how they interpreted or used their symbols of ethnic identity. For instance, Chineseness in Canada was strongly associated with working in STEM fields. While it is tempting to put this down to the different main industries of both cities, London also has a large academic and R&D sector, while many of my Toronto informants were not in STEM subjects (or had shifted to other professional areas after beginning in STEM). Furthermore, the multivalency of symbols meant that shared symbols of identity took on different connotations in different contexts; in the case of Taiwanese food, for instance, London's discourses were framed in the context of the city's gourmet culture, and Toronto's in terms of a general fondness for fusion cuisine, which mixed Asian and European culinary elements. This makes sense in light of the two cities' different global status: London approaches "foreign" food as part of a cosmopolitan, sophisticated search for "authenticity," while Toronto's experience is framed around a normalization of the migrant experience, hence the Night Market sponsored by the T&T grocery chain. While expatriates and other elite transnational migrants have been studied a number of times in different global cities (for instance, by Ryan & Mulholland, 2014; Teo, 2011; Yeoh & Willis, 2011; and others), a comparative approach shows how the same symbols can be used to construct subtly different discourses of identity in different settings.

The flexibility of identity and the symbols used to express it also allowed for connections with different groups. For instance, holidays such as the Mid-Autumn Festival could be used to indicate a connection with mainland China or with a pan-Asian identity (for many other Asian countries also celebrate it). Some argued that a fondness for food was a Chinese rather than specifically Taiwanese trait. In Toronto, the Night Market and (Taiwanese Canadian–owned) T&T (Ebner, 2009) were both pan-Asian in terms of the foods they offered. Likewise, the association of Taiwaneseness with food was different in Taipei, where interest in international cuisine was more normative and local foods showed a definite Japanese influence, whereas the London community, according to Glinda, was retreating into a more traditional and reified concept of Taiwanese food (see also Cappellini & Yen, 2013). For people employing *waishengren* discourses of Taiwaneseness, a geographic origin *outside* Taiwan could symbolize Taiwanese identity as much as an origin within it. In this regard, Wade-Giles romanization and traditional Mandarin characters are not uncommon in sections of the Chinese diaspora more generally, as well as in Hong Kong.

Finally, there was the case of *guanxi* networking. Such networks did exist, being referenced both explicitly and, when people spoke of "doing favours," implicitly. However, it was not always seen in the positive light shown in the literature (Park & Luo, 2001). For example, when Frankie talked about *guanxi* networking with mainlanders, it was less as an asset and more as part of a discourse about how mainlanders are less professional and rigorous than Taiwanese or Europeans. It could also be complicated by other forms of networking: it was ambiguous how much of Mrs Wang's influence in the alumni networks was due to her participation as an alumni association officer, how much was down to *guanxi* (as she did speak of her connections in terms of favours and exchanges), and how much to a kind of extrafamilial bamboo networking that involved developing fictive kinship relationships with fellow alumni. While *guanxi* is important, it is part of a general construction of useful global networks, not so much something that has always been uniquely indicative of Chineseness. The complexity and multivalency of symbols therefore affected the construction of networks, providing avenues for, and barriers to, networking in sometimes unexpected ways.

Complex Identities and the Network Society

The case of the Taiwanese skilled labour migrants shows how individuals present their identities strategically in order to further their own interests while also, in the process of negotiating the construction of their identities, contributing to wider discourses. Taiwanese skilled labour migrants present themselves differently in different contexts, using the same set of symbols: interpreted differently, a symbol such as Chinese food or full-form characters could be used to express an identity as Taiwanese, as Chinese, as a migrant, or – in certain ways relating to the cosmopolitanism of such symbols – as pan-Asian, as Canadian, or as a Londoner. However, as per Alvesson, Ashcraft, and Thomas (2008), other people's definitions also had to be taken into account, for instance, by not correcting people outside the community who lump "Chinese" people into a single category. One interviewee, Ruth, had no personal connection with China and self-identified as British while acknowledging that, since others see her as Chinese, she has had to accept a connection with this identity. Identity is thus performative but also discursive in this transnational context, and the performative aspect is in a mutual support relationship with the discursive aspect.

The discourses had connections to wider power relations outside of the immediate expression of identity. For instance, Taiwanese people do make use of "Chinese" community associations, but in doing so, they are tacitly reinforcing questions about what defines Chinese identity, the extent to which Taiwanese identity can be distinctive in its own right rather than a part of the Chinese diaspora, and the relative positions of different groups, or languages, or sets of characters, in the local Chinese community. Looking at multiple groups that can lay claim to the same set of identities thus expands on studies of single groups, in showing how these performances play out in different ways in different contexts and how all of these different performances are themselves connected through discourse.

In terms of studying global identities more generally, it is worth noting that, while they may have some common points – being multisited, overarching, and engaged with the drivers of globalization – they are also seen through different lenses at different sites. Social tools for crossing borders also present different choices: networking organizations, professional development, university attendance, family networks or networks built of obligations (here, bamboo and *guanxi* networks), and the use of online media also shape globalizing activities. Nonini and Ong (1997) write that viewing globalization from an Asian standpoint provides a perspective quite different from the North American and European ones. Studies of "cosmopolitans" (Iyer, 2000; Appiah, 1998), or "the transnational capitalist class" (Sklair, 2001), or "global youth" (Cappellini & Yen, 2013), or other forms of global identity have to take into account the fact that there are different globalizations, and differences between global cities and between global networks, and that these have a significant impact on the type and nature of global engagement that people can exploit.

While Castells's basic principles still more or less hold firm – even down to Castells anticipating the current protectionist, terrorist, and anti-globalization discourses (1997/2004, ch. 2) – the context of globalization is not precisely the same as at the turn of the millennium. While my data predate Brexit and the decline of the United States as a global power, some of the participants were clearly anticipating greater barriers to global activity; for instance, the Taiwanese Chamber of Commerce in London was developing networks with the explicit aim of combating local protectionism, and many of my interviewees in all three locations questioned how long China would remain the dominant power in Asia. Also, a significant number of my interviewees had migrated *before* the rise of present-day globalization, in the 1960s and 1970s, but still engaged in global networking at the time. (Vivian told a long anecdote about how she, as a bored faculty wife in an isolated American town in the 1970s, had introduced the local population to "authentic" Chinese cuisine, contributing, as it were, to cultural globalization.) This study can develop earlier research on globalization by showing that globalizing activities continue even at times, and in societies, where such activities are not part of the mainstream or considered generally positive.

Looking at the Taiwanese from a multisited perspective has not only upheld Castells's basic principles of identity-based social activities along global networks between nodes at global cities, but also developed it by considering that globalization is not always qualitatively the same at each node and that the context of globalization changes. It also shows that globalizing activities continue, even outside of formally identified "periods of globalization."

Differentiated Globalization: A New Approach to the Network Society

The image of the network society that has emerged from the Taiwanese perspective is of an intersection of network(s) and location(s), with individuals existing as nexus points of different local and global networks and particular locations (see Figure 9.1). Identity, expressed through symbols, can be used to navigate this complex web, because the multivalency of symbols means that the Taiwanese can link to different groups at the same time; the same symbols can cross borders and be recontextualized in different locations. I would argue that the Taiwanese experience, at least, should be seen in terms of "differentiated globalization": a complex pattern in which individuals operate simultaneously not just in the local and the global, but in multiple global and local contexts, to different degrees.

Figure 9.1

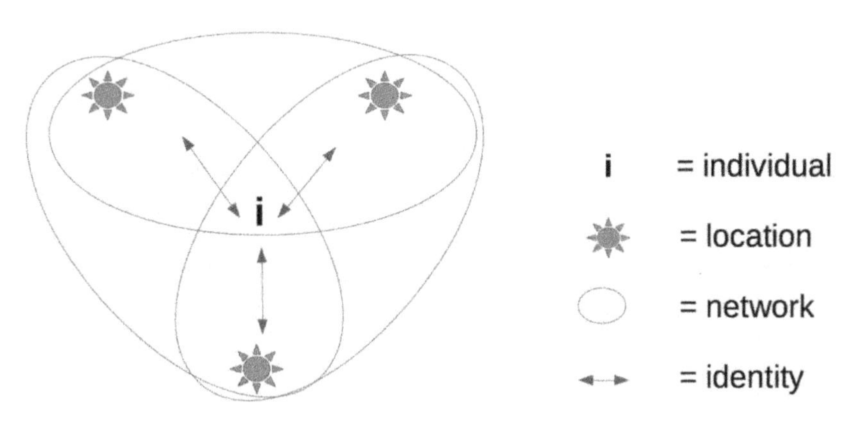

Contested Identities and Globalization(s)

A key point that emerged involved the concept of identity as simultaneously national and transnational. Glinda, during our interview, argued that while the mainland government attempted to link *Zhonghua* with *Zhongguo*, it were not necessarily the final arbiter in this situation; she was implying that its effort was part of wider discourses within the Chinese diaspora, and indeed outside it. This harmonized with the fact that Wendy spoke of making more friendships among diaspora Chinese than among mainlanders, implicitly through the common experience of being Chinese but not in the *Zhongguo* sense or, given the British context of these friendships, of being Chinese abroad. One might also consider the efforts of the KMT to argue in favour of a global Chinese identity beyond the mainland (Yu & Bairner, 2008; Murray & Hong, 1991). Informants of around my age indicated that during the dictatorship, children were taught that Mongolia, Hong Kong, and Tibet were part of a greater "free Chinese" identity group (see Liu & Hung, 2002). There was also the globalizing Sinophilia seen in the context of such organizations as Xinbeitou 8 and events such as the Taiwanese food festival, in which non-Chinese could participate in Chineseness through their appreciation of the language, food, or culture or through their experiences of having lived and worked in China or Taiwan. While this may not be the expression of a Chinese identity, it is the expression of an identity that has a connection to Chineseness. This has also been observed by Ong (1999), who writes that the forces of globalization have created a "transnational Chinese public" by allowing ethnically Chinese people to connect with Chinese culture wherever they are. The Taiwanese interpretation of Chineseness has thus added to the construction of Chinese identity as a global rather than nation-based ethnic identity.

This situation is not unique to the Chinese; it can also be seen in the context of European identities. The Germans, for instance, are often said to have a strong sense of local identity tied to a particular geographical region (Weigelt, 1984, Greverus, 1979) and are not generally considered, these days, as having a powerful group identity as a diaspora abroad. Yet studies of German expatriates indicate that German identity in this context, while it may have connections to a particular geographical construct, is as much a state of mind as anything else (Moore, 2005; Helbling & Teney, 2015). In an earlier study, I considered that the concept of *Heimat*, or "homeland," which is normally considered a quite geographically specific idea (see Applegate, 1990), becomes much more flexible for Germans living abroad (Moore, 2005, pp. 43, 125–126). The Taiwanese/Chinese debate also reflects, in inverted form, Chapman, Clegg, and Gajewska-De Mattos's (2004) observations of German and Polish identities along those two countries' shared border: each group acknowledges its shared traits as well as, implicitly, the fact that the border is something of an accident of history rather than a genuine line of cultural difference; at the same time, both reject the idea that they might have an identity in common (see also Dürrschmidt, 2006). While the Chinese experience may be unique in itself, it is also in line with what has been observed in other transnational ethnic communities, and it should not be considered as divorced from any other experience. This study, taken with other studies of transnational migrants such as the Germans, supports the idea that national identities may be less a matter of being from a particular location, than a kind of globalized identification with a particular group.

Furthermore, the denationalized, transnationalized nature of ethnic identities lends them a flexibility that allows their members to link to other identities. Ong's (1999) exploration of transnational Chinese identities has led her to consider the development of a transnational, pan-Asian identity through Chinese connections with other countries such as Singapore, Malaysia, Vietnam, and, indeed, Taiwan. Funnell (2014) notes the embracing of a transnational ethos in popular *wuxia* film, linked to a global Chineseness (which even extends out to stars who are not, in fact, ethnically Chinese, but Asian or European), which itself leads out to wider careers in the global film industry. The present study also highlights the different interconnected processes that encourage the formation and continuation of transnational communities.

In cases of contested identity, however, such as that of Taiwan, the debate over what this identity constituted, added to this development of nation-connected but transnational identity, leads to complicated symbolic conflicts. Paradoxically, following Ong's (1999) view, a discourse of Taiwan as "pan-Asian" can support the discourse of Taiwan as "Chinese," just as much as the discourse of Taiwanese identity as Chinese identity can support (Yang & Chang, 2010), subvert (Yu & Bairner, 2008), or explicitly challenge (Harrison, 2007) mainland China's discourses of Chinese identity. The Chinese diaspora generally can be simultaneously Chinese and not-Chinese. Having transcended individual nations, Chinese identity is thus subject to a transnational contestation, played out in a variety of different locations, as to what it constitutes; it stands to reason that similar discourses are being played out, in less dramatic fashion, with other national identities. The concept of identity, in transnational contexts, reflects and shapes political and economic activity in local areas.

The means by which national boundaries are crossed and transcended also have this complex, discursive aspect. For instance, as much as Chinese and Taiwanese identities, and the discourses surrounding them, transcended national borders, they were also influenced by national governments, a paradox that has been a key theme of transnationalism from its outset (see Hirst & Thompson, 1996; Tomlinson, 1999). The Taiwanese government supports and runs student associations and is indirectly linked with other networking organizations; the migration policies of the British and Canadian governments, and their political discourses of who is an acceptable migrant, affect the ways in which the Taiwanese people are able to relate to the mainland, with, as noted, the British cohort approaching the migration experience as temporary and limited, and the Canadian cohort planning to settle and not discourage their children from assimilating. Furthermore, there is the situation mentioned earlier regarding the chambers of commerce, which receive political support from the Government of Taiwan but are also technically independent, which well places them to broker alliances by encouraging business ventures even in countries that do not officially acknowledge Taiwan. The fact that Taiwan is not recognized officially as independent by either the UK or Canada, even though businesses from either country will deal with Taiwanese companies as independent entities, indicates the complexity of such connections. A discourse that transcends national borders is thus sometimes played out through national-level policies and institutions.

Furthermore, the people in this study were usually embedded in many globalizing organizations or processes simultaneously. A single individual could be involved in a number of active networking, social, and professional organizations, such as alumni groups, student groups, chambers of commerce, and business networks, as well as more informal and diffuse networks such as *guanxi* networks and bamboo networks, or even simply maintain a social media presence. All of the above groups have different elements of globalization and localization as well as diverse connections to a number of localities with varying degrees of strength. Individuals' global presence, in terms of activities through which they express identities, thus form a complicated web as those identities are expressed through many different media, in many different social contexts.

Differentiated Globalization

At this point, it might be worth considering Glick Schiller and Salazar's (2012) take on mobility as a dynamic between rooted and rootless social movements. Applying this to national and ethnic identities, this fits with Chen (2012)'s argument about Taiwanese identity – that, while many commentators view the possession of Taiwanese and Chinese identities as sequential (i.e., held one at a time depending on context), in fact, they can be held at the same time. Diasporas and labour migrants are as much transnational social formations as multinational corporations are, so it may be useful to consider them in the same way, as complicated, networked entities that simultaneously incorporate conflicting impulses and directions (see Dicken, 2010). What appears on one level to be a clear albeit transnational entity cannot be understood as simply that. If ethnic identity is complex and transcends geography, then so must be the organizations that allow its expression.

Furthermore, considering these activities in historical perspective, we may be looking at a formulation that predates the use of nations as administrative units and focuses of nationalism. One might, for instance, consider Taiwan's problematic relationship with the mainland; while the *waishengren/benshengren* divide and the mainland's claim over Taiwan may throw that relationship into focus, the mainland's interest in claiming Taiwan as Chinese has varied dramatically in intensity and practice since the Middle Ages, and Taiwan, meanwhile, views Japan and several European countries as fundamental to its historical past as a colony. The idea that nations are fuzzy, identity-defined entities that may not have a direct relationship with a particular territory is clearly nothing new, and may, arguably, reflect the historical norm.

Also, it is worth emphasizing again that labour migrants are not just historically embedded and influenced by recent events; rather, different periods of globalization inform one another. Present-day Taiwanese globalization has been influenced by globalizing events that occurred in the seventeenth and late nineteenth centuries. Moreover, the present period of globalization has not been uniform: while it is generally agreed that a state of globalization has existed since at least the 1980s (see Held et al., 1999, it is in quite a different place in the 2010s than it was in the 1980s, the 1990s, and the early 2000s. Finally, globalizing activities also occur during eras that are not typically viewed as periods of globalization: the Cold War may not have been a time when technological advances and weakening social barriers meant that "the constraints of geography on social ... arrangements recede" (Waters, 1995, p. 3); but transnational activities nonetheless took place then – diasporas were established, multinational corporations developed, and communications systems became more and more able to cross borders in real time; without all of that, the post–Cold War period of globalization could not have happened. It is possible that, instead of thinking in terms of eras of "globalization," we need to see globalization as a process that is always present to some degree, but in greater or lesser degrees depending on the social context and the technological/social impulses that drive it.

Transnational Social Networking

The present study has also explored the wider implications of the networking aspect of globalization. In studies of international businesses and skilled labour migration, this has largely been considered in terms of the literature on global production networks (Coe, Dicken, & Hess, 2008; Dicken, 2010); elsewhere, researchers have considered diasporas (Ho, 2011a), or individual social networks (Montgomery, 2008), or the construction of elite identity through membership in certain professions or attendance at particular universities (Li, 1994), as examples of transnationally networked social formations. Glick Schiller and Salazar (2012) argue that mobility, more than stasis, has been the historic norm, notwithstanding that the academic focus has shifted only recently to migration. In this sense, we are not so much in a post-national as in a neo-national world, one in which nations exist but their congruence with a particular territory (as in Smith, 1995; Eriksen, 1993) is no longer cut and dried, and where some nations may indeed have global extensions.

In addition, these identities are not simply reinterpreted in different contexts – they are *recontextualized* (see Brannen, Liker, & Fruin, 1999; Brannen, 2004). Recontextualization is the process whereby different symbols can, while remaining externally the same, take on different meanings and referents in new cultural contexts. The same symbol, with the same *ostensible* meaning, changes from context to context in terms of its connotations and associations. In the context of this study, I would go further and note that the symbols through which people express their identities as Taiwanese are the same, or similar, in all the different contexts, and are used in analogous ways to build both internal and external networks. However, the multivalency of the symbols means they can be recontextualized in line with the various local discourses, in such a way that ostensibly unified and closed ethnic networks lead to different, seemingly unrelated identities. In a transnational context, this allows for greater opportunities to integrate, and to develop local and transnational connections, than a simpler and more straightforward use of symbols would.

Furthermore, these "transnational nations" are in no way unified, even, and possessed of universal values. Taiwanese migrants may participate in "global Chineseness," but their individual definitions of what this means, and how they see or present themselves as Chinese, may be very different from a mainlander's. A particular "global city" may be simultaneously local and transnational in its affiliations (see Rogers, 2005; Yeoh, 2005), and in different ways than other global cities. Much as Krätke (2014) identifies "multiple globalizations" based on differences between business sectors and their different levels and types of globalization, and Glick Schiller and Salazar (2012) speak in terms of "regimes of mobility," so we must recognize that these multiple globalizations inhabit different terrains and social maps, even before we begin to consider the impact of the different globalizations of particular "global cities" and the complications created by the intersections of different globalizations.

Based on all this, I contend that we need to consider globalization not as "uneven" or as "multiple" so much as "differentiated." We are not looking at several globalizations distinct from one another, with clear boundary lines; nor are we looking at a situation where some groups and places are more local or global than others, as groups and locations may sometimes be equally global, but in different ways. So it is helpful to consider globalization as a quality or as a state of being that may be present to a greater or lesser extent, depending on the circumstances. The concept of *differentiated globalization* can therefore be used, beyond the confines of the Taiwanese study, to signify the complexity of social organization in transnational social spaces, with different forms, types, and orientations of globalization interacting and participating with one another, to different degrees, in a complex, ongoing discourse.

The differences of globalization have been considered from the early days of research on the subject, in the literature on class and globalization. For example, Portes (1998) and Guarnizo and Smith (1998) have considered whether subaltern labour migration is categorically different from skilled labour migration. These studies explore the complexity of identity in transnational contexts, thereby problematizing the categories of "elite" and "subaltern" (Portes, 1998), a social complexity and multivalency that is in line with the present study. Those studies, however, conclude that working-class migrants employ the same globalizing socio-economic forces as the elites, in order to further working-class aims, whereas the Taiwanese case suggests a situation in which different globalizing and localizing opportunities influence migration: the aims of the chambers of commerce, and their reach, are similar to those of the alumni associations, but also different from these, and this affects how their members use them as well as the opportunities they present to their members. To understand transnational social activities, we must consider differences as well as similarities in globalization. The evidence is that many groups are global, but all are *differently* global.

The Taiwanese case also threw up some interesting paradoxes. For instance, although the general discourse in London was becoming protectionist and isolationist, while the discourse in Toronto was more cosmopolitan, welcoming immigrants and celebrating difference, the result was to encourage more local connections in the Toronto case, and more global ones in the London case. Taiwanese people in Toronto had settled, developed stakes in the community, and encouraged their children to integrate, whereas the less friendly attitudes in London drove Taiwanese people to at least think in terms of future migration and to encourage their children to keep up their global connections. Maintaining a positive or negative attitude toward globalization and migration, therefore, doesn't necessarily have the anticipated effect.

The other area of complexity that can be highlighted here has to do with how ethnic identity is actually used. For the Taiwanese, identities are simultaneously local (in this case, national) and global, and this can lead off into other identities. It seems that identity here is not just uneven but inherently multiple. It is not simply that there are uneven connections between global and local; there are also different globalizations, which are experienced in different ways: through identity, through the organizations that allow the expression of identity and its development, and through the differences between global cities, global organizations, and global networks. Finally, we also have to consider the historical interconnectedness of globalization events, for earlier periods of globalization inform later ones. The case of a single category of skilled labour migrants, linked with a particular location (Taiwan), seen across different locations and in different social contexts, shows globalization to be a differentiated phenomenon that is expressed in many ways in many places and times.

Implications and Future Directions

Implications for Research

The study of Taiwanese skilled labour migrants suggests a number of directions for future research. To begin with, it supports earlier studies of the ways in which seemingly localized or national identities can be used to build transnational communities (see Moore, 2005; Helbling & Teney, 2015). In this case, we have a global Chineseness that not only does not require one to be in, or come from, China, but also can serve as a counter-narrative to mainland China's claim to a normative Chinese identity.

This study provides a critique of, and alternative to, reified studies of labour migration that employ the nation as a unit of culture. As noted by Chapman (1997) and, more recently, by Tsui (2016), such studies tend to emerge from a behaviourist and quantitative background, leading to a functionalist and reductive approach to social and cultural phenomena, as well as to the persistence of the ecological fallacy that assumes that nations are *de facto* units of culture (see McSweeney, 2002, 2013). Consequently, studies of Taiwanese business tend to efface precise issues such as the *waishengren/benshengren* distinction, or differences made by age and ethnicity. Furthermore, studies of expatriates often tend to treat them as if they were undifferentiated units of "cultural programming": that an expatriate from, for instance, China, is a unit of Chinese culture, uninfluenced from the outside. This ignores the fact that many expatriates may well have studied abroad or taken multiple assignments overseas; even if not, they may well have family abroad, have travelled, or, simply, absorbed the popular culture of such countries as the United States, Japan, and the UK. As this study has shown, trans- or meta-national identities such as global Chineseness may also play a role – one that, again, is less often acknowledged in the business literature. And even when it is acknowledged, it seems to be wrongly assumed that it denotes a monolithic group that is unified in its beliefs and values. Taking a functionalist approach to culture erases the internal divisions within societies as well as the existence of meta-national cultures; a more reflexive approach provides a more useful way of addressing transnational social formations in business.

This does not, however, entirely do away with the concept of the national. People clearly identify in national terms, and the nation is not without social meaning – as, for instance, in Kevin's preference for Taiwanese over European suppliers (similar to that shown by the Taiwanese expatriate managers studied by Chen [2004]) – suggesting a national identification that goes beyond practical distinctions, or that *guanxi* networks can be nation-based. One reason why some people view globalization as under threat is the rise of nationalist discourses. However, some of the complications noted above might be resolved by taking an identity-based approach to nations: thinking of them less in terms of national culture than in terms of an identification with one or more national cultures, and more as a social identity than as a particular set of traits linked to a particular territory. As Castells noted, nationalist discourses can, paradoxically, encourage globalization.

Finally, this study raises a few points about the relationship of research to practice, and researcher to subject. Much the same uses of identity and network are noted among the Taiwanese (and, more generally, the Chinese) as among the Germans, yet the literature on the Chinese tends to focus on "diaspora" (Ho, 2011a) or on *guanxi* and bamboo networks (Xin & Pearce, 1996; Tien & Luan, 2015). Numerous studies have discussed the impact of Confucianism on Asian business and management (e.g., Wong, Shaw, & Ng, 2010), but there is no similar strand of literature considering the comparable impact of Christianity, as a social and philosophical movement, on European corporations. This suggests a degree of exoticization, arguably orientalism, in the social sciences, in which activities common to most, if not all, transnational labour migrants are exoticized in an Asian context but not in a European one. As noted elsewhere, the existence of large communities of overseas Canadians has gone largely unconsidered in the social sciences, and the documented existence of practices similar to *guanxi* among Europeans is also little discussed (see Xin & Pearce, 1996, p. 1642); the same goes with analogous concepts in other Asian countries (see Alston, 1989). Indeed, at least one study suggests that the "bamboo network" may be overrated (Tien & Luan, 2015). Rather than focusing on differentiations, it might be better to consider networks in context and how their similarities and differences to others allow greater understanding of transnational social activities.

Implications for Policy

This study has shown clearly that migration policy does not simply stem from the political environments that produce it; it also influences local and transnational culture. A comparison of the UK and Canada provides a useful illustration. The migration policies of both countries have clearly influenced the settler-focused and sojourner-focused discourses in both locations, as well as the ways in which the Taiwanese labour migrants relate to their social and political environments. However, they also reflect more than simply local attitudes toward migration: the idea that London is a good place to live for a few years, but not for the long term, exists among local people as well as migrants, and it can be argued that Canada's discourses of migration and settlement are rooted in its past as a colonial nation. The two-way relationship between policy and culture needs to be more closely considered.

Moreover, attitudes and policies change over time. Koopmans and Statham's (1999) study of migrants in Britain and Germany in the twentieth century found that migrants in the UK engaged in local politics to a greater degree than their German counterparts due to narratives of inclusion in the UK: the fact that, in the 1950s, Britishness was seen as something that could be "acquired," rather than as something into which one was born, affected migrants' perceptions of whether they would be able to engage with the community. However, this argument somewhat contrasts with this study's findings about the present-day experiences of Taiwanese migrants to the UK: they largely saw themselves as having to work around political processes, not as engaging with them. It is unclear whether a particular national or ethnic identity can change over time, or be "acquired." Furthermore, in line with the earlier discussion of reflexivity, it is also worth noting that research affects both policy and culture: the formulation of policy, and its interpretation and engagement by skilled labour migrants, both happen in the context of educated, aware professionals who have access to the academic literature on the subject. More reflexive attention thus needs to be paid to policy and legal structures, in particular to how they both emerge from and influence the wider culture.

It is also worth considering how national and regional migration policies can utilize the concept of differentiated globalization. The most obvious manifestation of engagement with skilled labour migrants has been individual nations treating their diasporas administratively as an extension of the country, taxing them, and allowing them voting rights and other domestic privileges (Ho, 2011a), but it is also evident, in the case at hand, in the Taiwanese government's political interest in its student associations. However, such policies again assume an identity related to a grounded, national location, and do not really take into account the multiple affiliations of transnational skilled labour migrants. The engagement of online and technologically mediated citizenship may well suggest a similar new direction toward the development of global identities. However, we may also be seeing a simultaneous movement toward increasing *local focus*, as transnational tariff agreements and political unions are increasingly questioned, and *global reach*, as people and companies nonetheless continue to cross borders – indeed, they may be doing so at greater rates in pursuit of social, economic, and human resources. So the form a post-national, or meta-national, approach to policy might take is uncertain.

Directions for Further Research

This study, inevitably, has gaps and lacunae that point toward further directions for research. The samples of interviewees, for instance, are numerically uneven, based simply on the opportunities that came available to me in each location (it was, for instance, much easier for me to interview London-based informants than those in the other two locations). Also, there was usually more in the way of ongoing informal social contacts with the Toronto interviewees, owing to the intersection of that particular network with my own friends and family; arguably, this balanced out the smaller interview sample size. In Taipei, moreover, the primary purpose of the visit was to conduct background research and engage in participant-observation with a view to understanding where my informants were coming from. However, I will reiterate that my results are not to be taken as a blanket representation of all overseas Taiwanese, even in the studied locations; they are more by way of working toward understanding the transnational social formations in which they participate, through analysing the experiences of the members of those formations in different ways.

It should also be repeated that this study has been filtered through the experiences of a single researcher, albeit with the vital assistance of many helpers and informants, and that, while this presents particular opportunities, it can also close off certain others. Alternative perspectives have been sought out through interviews and references to the literature. Additionally, the longitudinal nature of the study – six years old at the time of writing – has assisted with developing different perspectives, for instance, as Taiwan moves from a KMT-dominated political establishment to a more balanced one, and as the UK's isolationism deepens while Canada moves from an extreme right-wing government toward a more socialist model. Further research will allow the different communities to be revisited over the next few years, and earlier conclusions will undoubtedly be revised and developed.

Finally, it might be worth considering the same issues discussed here through different analytical lenses. This study focused on a group defined primarily by ethnic identity and explored the ways in which that identity extended out into others such as professional, academic, and meta-national. It might be worth considering, instead, a group defined by professional identity, with extensions into other identities such as ethnic ones, and exploring whether such a shift in perspective makes a significant difference. It is also worth considering other meta-identities than the one considered here: a number of scholars, for instance, view queer identity as a globalizing one, and sexual orientation as meta-identity presents interesting research possibilities (see Cruz-Malave, 2002). It might also be worth adding an element of organizational identity (see Albert & Whetten, 1985) and exploring how differentiated transnationalism looks from an organizational perspective, taking the global production network model and adding to it the unevennesses, different engagements, and simultaneous contradictions of the differentiated model as part of the mix. Further research on identity might thus develop the conclusions of this study.

Future Considerations

While it is impossible to predict the future, it might be worth considering some of the developments that might emerge from the situation described here. Anti-migrant and anti-EU sentiment in the UK, and similar political developments elsewhere, might suggest that the era of free movement of people and capital across borders is threatened (Redclift & Rajina, 2019; see also Stratham, 2003). While Canada remains overall more tolerant, there has been a rise in recent years of rhetoric, and indeed of policy, that is less welcoming to migrants (Vomiero & Russell, 2019). Also, a prohibitive rise in the cost of transportation and a dearth of key resources could signal an end to (or, at least, a diminuition of) globalization and a reduction in transnational activity. However, this is debatable; the "War on Terror," which might have sparked a closing down of borders because of its increased focus on national security, instead has led to an era of transnational surveillance and of awareness of cross-border criminal activities. Cyberattacks and the terrorist activities of extremist groups are firmly embedded in the transnational (Kshetri & Alcantara, 2015). Likewise, the recession of 2008 has, in some areas, led to a deepening rather than a lessening of transnational activities (van Veen, 2018) as companies seek to cut costs by pursuing cheaper labour forces or by engaging in technologically mediated cross-border management rather than physically sending expatriates. Also, the global coronavirus pandemic has seen people substituting virtual media for travel rather than ending transnational activities entirely. In such cases, the closing down of national borders has led to an increased awareness of meta-national activities, with, perhaps, national identities actually weakening in favour of regional, local, and ethnic ones. Globalization may be changing, but it is likely wrong to argue that it is ending.

Furthermore, as discussed above, even an end to the period of globalization does not mean the curtailment of transnational activities. Individuals will still cross borders and exchange information and goods around the world; it is worth noting that despite the different policies and cultures in the UK and Canada, both cohorts included sojourners *and* settlers, indicating that having policies and cultures hostile to migrants will not necessarily prevent migration (nor will having policies and cultures friendly to migrants inherently encourage them). A differentiated, historically embedded perspective on globalization is therefore needed to take into account the changes emerging in the modern world.

Conclusions

The material in the preceding chapters was presented with a view to addressing the question of how skilled labour migrants use their ethnic, and other, identities to construct and use transnational social and business networks. Taiwanese transnational professionals at three sites around the world were studied: two diaspora groups, one in Europe and one in North America, in addition to an ethnographic visit to Taiwan itself. Three analytical lenses were used to consider the data: (1) location, (2) ethnic networks, (3) networking organizations. The study noted the differences each lens made, how they influenced one another, and how ethnic identities, in each context, allowed connections through multivalent symbols with other locations, networks, organizations, and identities. The study concluded, first, that the Taiwanese transnational professionals use the multivalent properties of symbols to link their ethnic identities, through symbolic self-presentation, into others – professional, academic, and also, in this case, other ethnic or "meta-ethnic" (pan-Asian, global Chinese) identities. Second, the multilayered and discursive aspects of this process mean that contrasting and even contradictory discourses of identity can be going on simultaneously on different levels (for instance, being at once Taiwanese and Chinese, or being globally connected but locally embedded). Finally, the study noted that new approaches to the study of identity and/or transnationalism need to be developed so as to take this multiplicity and simultaneity into account.

This study, therefore, leads out to a number of directions for new research stemming from its conclusions. In the first place, it highlights that identity-based approaches to transnational groups are needed. Furthermore, those approaches need to be complex and multisited rather than situated and contextualized, and they must also take into account the historical circumstances underpinning globalization, transnationalism, and migration. Globalization should be seen less as "uneven" than as "differentiated": while some places are indeed "more global" than others, it may also be that places, groups, and activities are in fact *differently* globalized. Finally, the multivalency of symbols mediates and encourages the navigation of transnational social spaces and thus the potential for developing connections with other identities and for recontextualization in different locations. Such symbols allow individuals and groups to navigate the complexities of location and of networking across borders.

This study also suggests that social scientists in fields related to globalization need to work on developing multisited, or multisiting-friendly, as it were, methodologies and studies. New approaches to globalization are needed that acknowledge location effects and that do not presuppose undifferentiated social systems. Reflexive approaches are also increasingly useful, to take into account the connections between research, policy, practice, and different cultures.

Finally, the differentiated globalization approach, when applied to understand the experiences of Taiwanese skilled labour migrants, suggests that we need to consider transnational activities as a function of connections between network(s) and location(s), mediated by symbolic expressions of identity, and as historically and socially embedded. We also need to take into account the emergence of particular sorts of "metanational" identities that are nation-based, and differentiated, but also transnational and identification-focused. In this way, a network-based but location-sensitive perspective on transnational social formations can be developed.

Appendix One

List of Interviewees

UK Cohort

Name	Gender	Ethnic self-identification	Age	Employment	Source of contact
Rupert	M	Taiwanese	70s	Accountant (ret.)	World Taiwanese Chambers of Commerce
Wendy	F	Taiwanese	20s	NGO employee	Taiwanese Students Association
Michael	M	Taiwanese	30s	Lawyer	WTCC
Steven	M	Taiwanese	40s	Civil servant	WTCC/Taipei Representative Office
Dougal	M	Taiwanese	40s	Civil servant	TAITRA
Penelope	F	Taiwanese	70	Teacher (ret)	Buddhist temple/Food Festival
Paul	M	Taiwanese	50	Deputy managing director (electronics company)	University
Lali	F	Taiwanese	40s	Entrepreneur	Food Festival
Trevor	M	Taiwanese	50s	Company director	WTCC
Hua	F	Taiwanese	70s	Teacher	WTCC
Maria	F	Taiwanese	50s	Entrepreneur/travel agent	WTCC
James	M	Taiwanese	20s	Postgraduate student	Oxford Taiwanese Society
John	M	Taiwanese	20s	Postgraduate student	Oxford Taiwanese Society
Daniel	M	Taiwanese	20s	Medical student	Taiwanese Student Association
Mr Lee	M	Taiwanese	40s	Civil servant	Representative Office

(*Continued*)

(*Continued*)

Name	Gender	Ethnic self-identification	Age	Employment	Source of contact
Mr Hsiao	M	Taiwanese	60s	Civil servant	Representative Office
Carla	F	Taiwanese	50s	Civil Servant	Representative Office
Ruth	F	British (ethnically Chinese)	20s	Student	University
Stan	M	British (ethnically Chinese)	20s/ 30s	Councillor	University
Franklin	M	Chinese	50s	Community centre director	Camden Chinese Community Centre
Cassandra	F	Australian (ethnically Chinese)	20s	Lawyer (trainee)	University
Simon	M	Chinese	20s	Community group leader/online manager	University
Carrie	F	Chinese	20s	Community group leader	University
Caroline	F	Chinese	40s	University lecturer	University
Blake	M	Hakka Taiwanese		IT manager	WTCC
Glinda	F	Taiwanese	40s	University lecturer	University

Canada Cohort

Name	Gender	Ethnic self-identification	Age	Employment	Source of contact
Mrs Wang	F	Taiwanese (sometimes Chinese)	70s	Chemist and IT professional (retired)	Friends/family
Kay	M	Taiwanese/ pan-Asian	30s	IT professional	Friends/family
Linda Hua	F	Taiwanese		Speech pathologist	Friends/family
Lucy	F	Taiwanese	30s	Architect	Friends/family
Lucille	F	Taiwanese		Real estate agent	National Taiwan University Alumni
Nora	F	Taiwanese	50s	Retired	NTU
Johnny	M	Taiwanese	70s	Engineer (retired)	NTU

Name	Gender	Ethnic self-identification	Age	Employment	Source of contact
Diana Chen	F	Taiwanese		Real estate agent	NTU
Bruce	M	Taiwanese	20s	Postgraduate student	Taiwanese Merchants Association of Toronto
Gordon	M	Canadian	60s	IT technician	Friends/family
Vivian	F	Taiwanese American	70s	Entrepreneur	NTU
Bill	M	Taiwanese American	70s	Retired; former chemist and manager	NTU

Taiwan Cohort

Name	Gender	Ethnic self-identification	Age	Employment	Source of contact
Frankie	M	Taiwanese	30s	Manager	NCCU
Sarah	F	Taiwanese	30s	IT manager for search engine company	NCCU
Marina	F	Taiwanese	40s	Online community manager for gaming company	NCCU
Jaime	F	Taiwanese	40s	Sales and marketing/ Entrepreneur	NCCU
Ms Wong	F	Taiwanese	50s	Entrepreneur	NCCU
Gloria	F	Taiwanese	30s	Manager, telecoms company	NCCU
Damien	M	Swiss	40s	Lawyer	NCCU
John	M	Taiwanese	20s	Graduate student	NCCU
Jim Hsiao	M	Taiwanese	50s	Entrepreneur	WTCC

Appendix Two

Indicative Questions from Semi-Structured Interviews

The purpose of the project is to study the way in which Taiwanese businesspeople construct transnational business networks, using their identities as Taiwanese and as professionals. All answers will remain confidential, and participants' identities will be disguised in any publications of the project's findings.

1 How would you describe yourself?
2 What does your current job involve?
3 What are your business connections with Europe? With North America?
3 How would you describe your ethnicity?
4 What would you say makes you this ethnicity?
5 What is it to "be Taiwanese"?
6 Would you also say you have a local identity? If so, what, and how does it affect your ethnicity?
7 How has your being from Taiwan, specifically, affected your work?
8 Can you tell me about any situations in which you felt your ethnic identity helped or hindered you at work?
9 Can you tell me about any situations in which you were better able to get information across/learn something new because of your ethnicity?
10 Can you tell me about any situations in which you found it more difficult to get information across/learn something new because of your ethnicity?
11 Have things changed for you in your work with the "rise of China" in the past ten years?
12 Have things changed for you in your work with the increased prominence of Taiwan as an Asian business centre in recent years?
13 Are there situations in which you play a "mediating role" between Chinese and non-Chinse people? Please describe them.

14 What, in your opinion, are the differences (if any) between a Taiwanese and a mainland Chinese manager? How much do they matter?
15 Are there any aspects of your work that you would describe as "global" or "globalizing"?
16 Are there any aspects of your work that you would say involve "knowledge" or "information"?
17 Can you tell me about any situations in which your ethnic identity affected the transfer of knowledge?
18 What sort of communications technologies do you use at work, how often, and how?
19 Do you feel your ethnic identity comes into play more in face-to-face interactions than electronically mediated ones (e.g. e-mail, Skype)? How?
20 Would you say your experiences are "typical" for people of your ethnicity/ethnic mix? Why/why not?
21 Are there any other questions I should have asked? Is there anyone else I should talk to?

Bibliography

Acuto, M. (2013). The geopolitical dimension. In M. Acuto and W. Steele (Eds.), *Global city challenges: Debating a concept, improving the practice* (pp. 170–187). Palgrave Macmillan.

Acuto, M., & Steele, W. (Eds.). (2013). *Global city challenges: Debating a concept, improving the practice.* Palgrave Macmillan.

Albert, S., & Whetten, D. (1985). Organizational identity. In L. L. Cummings and B. M. Staw (Eds.), *Research in Organizational Behaviour* (pp. 236–295). JAI Press.

Alinejad, D., Candidatu, L., Mevsimler, M., Minchilli, C., Ponzanesi, S., & van der Vlist, F. (2018). Diaspora and mapping methodologies: tracing transnational digital connections with "mattering maps." *Global Networks, 19*(1), 21–43. https://doi.org/10.1111/glob.12197

Alston, J. P. (1989). *Wa, guanxi* and *inhwa*: Managerial principles in Japan, China and Korea. *Business Horizons 32*(2), 16–31. https://doi.org/10.1016/s0007-6813(89)80007-2

Alvesson, M. (2002). *Understanding organizational culture.* Sage.

Alvesson, M., Ashcraft, K. & Thomas, R. (2008). Identity matters: Reflections on the construction of identity scholarship in organization studies. *Organization 15*, 5–28. https://doi.org/10.1177/1350508407084426

Amiraux, V. (1997). Turkish Islamic associations in Germany and the issue of European citizenship. In S. Vertovec & C. Peach (Eds.), *Islam in Europe: The politics of religion and community* (pp. 245–269). Macmillan.

Anderson, B. (2000). *Doing the dirty work? The global politics of domestic labour.* Zed Books.

Andrade, T. (2007). *How Taiwan became Chinese: Dutch, Spanish, and Han colonization in the seventeenth century.* Columbia University Press.

[Anonymous]. (2014). Canada's leadership in the extractive resource sector is helping to fight global poverty. *TendersInfo News*, 7 October 2014.

Appiah, K. A. (1998). Cosmopolitan patriots. In P. Cheah & B. Robbins (Eds.), *Cosmopolitics: Thinking and feeling beyond the nation* (pp. 91–114). University of Minneapolis Press.

Applegate, C. (1990). *A nation of provincials: The German idea of Heimat.* University of California Press.

Appleton, S. (1970). Taiwanese and mainlanders on Taiwan: A survey of student attitudes. *The China Quarterly, 44,* 38–65. https://doi.org/10.1017/s030574100004282x

Asato, W. (2017). Welfare regime and labour migration policy for elderly care: New phase of social development in Taiwan. *Asia Pacific Journal of Social Work and Development, 27*(3–4), 221–233. https://doi.org/10.1080/02185385.2017.1408489

Banks, M. (1992). *Organizing Jainism in India and England.* Clarendon Press.

Banks, M. (1998). Visual anthropology: Image, object and interpretation. In J. Prosser (Ed.), *Image-based research: A sourcebook for qualitative researchers* (pp. 9–23). Falmer Press.

Barber, P. G. (2008). The ideal immigrant?: Gendered class subjects in Philippine – Canada migration. *Third World Quarterly, 29*(7), 1265–1285. https://doi.org/10.1080/01436590802386385

Baumann, G. (1996). *Contesting culture: Discourses of ethnicity in multi-ethnic London.* Cambridge University Press.

BBC (2004). Bypassing China's net firewall. *BBC News Online,* 10 March 2004. http://news.bbc.co.uk/1/hi/technology/3548035.stm

BBC (2012). 2011 Census: 45% of Londoners white British. *BBC News Online,* 11 December 2012. https://www.bbc.co.uk/news/uk-england-london-20680565

Beaverstock, J. V. (1996). Subcontracting the accountant! Professional labour markets, migration and organisational networks in the global accountancy industry. *Environment and Planning, 28,* 303–326. https://doi.org/10.1068/a280303

Beaverstock, J. V. (2005). Transnational elites in the city: British highly skilled inter-company transferees in New York city's financial district. *Journal of Ethnic and Migration Studies, 31*(2), 245–268. https://doi.org/10.1080/1369183042000339918

Beaverstock, J. V. (2011). Servicing British expatriate "talent" in Singapore: Exploring ordinary transnationalism and the role of the "expatriate" club. *Journal of Ethnic and Migration Studies, 37*(5), 709–728. https://doi.org/10.1080/1369183x.2011.559714

Beaverstock, J. V. (2018). New insights in reproducing transnational corporate elites: The labour market intermediation of executive search in the pursuit of global talent in Singapore. *Global Networks, 18*(3), 500–522. https://doi.org/10.1111/glob.12196

Beaverstock, J. V., & Smith, J. (1996). Lending jobs to global cities: Skilled international labour migration, investment banking, and the City of London., *Urban Studies, 33*, 1377–1394. https://doi.org/10.1080/0042098966709

Bedford, O., & Hwang, K. (2006). *Taiwanese identity and democracy: The social psychology of Taiwan's 2004 elections.* Palgrave.

Beerepoot, N., & Lambregts, B. (2015). Competition in online job marketplaces: Towards a global labour market for outsourcing services? *Global Networks, 15*, 236–255. https://doi.org/10.1111/glob.12051

Bell, D., & de Shalit, A. (2011). *The spirit of cities: Why the identity of a city matters in the global age.* Princeton University Press.

Benn, C. (2004). Toronto. In G. Hallowell (Ed.), *The Oxford Companion to Canadian History.* Oxford University Press

Berghahn, M. (1988). *Continental Britons: German-Jewish refugees from Nazi Germany.* Berg.

Berry, J. W. (2005). Mutual attitudes among immigrants and ethnocultural groups in Canada. *International Journal of Intercultural Relations, 30*(6), 719–734. https://doi.org/10.1016/j.ijintrel.2006.06.004

Berry, J. W. (2011). Integration and multiculturalism: Ways towards social solidarity. *Papers on Social Representations, 20*, 2.1–2.21.

Björklund, U. (2003). Armenians of Athens and Istanbul: The Armenian diaspora and the "transnational" nation. *Global Networks, 3*(3), 337–354. https://doi.org/10.1111/1471-0374.00065

Black, D. (2015). Court challenge slams new Citizenship Act as anti-Canadian." *Toronto Star*, 20 August 2015.

Bloch, M. (1974). Symbols, song, dance, and features of articulation: Is religion an extreme form of traditional authority? *Archives Europaiennes de la Sociologie, 15*, 55–81. https://doi.org/10.1017/s0003975600002824

Bocker, A., & Havinga, T. (1997). *Asylum migration to the European Union: Patterns of origin and destination.* European Commission.

Borer, M. C. (1977). *The City of London: A History.* Constable and Co.

Brannen, M. Y. (2004). When Mickey loses face: Recontextualization, semantic fit, and the semiotics of foreignness. *Academy of Management Review, 29*, 593–616. https://doi.org/10.2307/20159073

Brannen, M. Y., Liker, J. K., & Fruin, W. M. (1999). Recontextualization and factory-to-factory knowledge transfer from Japan to the United States. In J. K. Liker (Ed.), *Remade in America: Transplanting and transforming Japanese management systems* (pp. 117–154). Oxford University Press.

Brannen, M. Y., Moore, F., & Mughan, T. (2013). Strategic ethnography and reinvigorating Tesco Plc: Leveraging multicultural teams using ethnographic method. *Ethnographic Praxis in Industry, 1*, 282–299. https://doi.org/10.1111/j.1559-8918.2013.00024.x

Brannen, M. Y., & Salk, J. (2000). Partnering across borders: Negotiating organizational culture in a German–Japanese joint venture. *Human Relations, 53*, 451–487. https://doi.org/10.1177/0018726700534001

Briggs, C. L. (1986). *Learning how to ask: A sociolinguistic appraisal of the role of the interview in social science research.* Cambridge University Press.

Brown, M. J. (2001). Reconstructing ethnicity: Recorded and remembered identity in Taiwan. *Ethnology, 40*(2), 153–164. https://doi.org/10.2307/3773928

Burns, T. (1992). *Erving Goffman.* Routledge.

Cappellini, B., & Yen, D. A. (2013). Little emperors in the UK: Acculturation and food over time. *Journal of Business Research, 66*(8), 968–974. https://doi.org/10.1016/j.jbusres.2011.12.019

Cappellini, B., & Yen, D. A. (2015). A space of one's own: Spatial and identity liminality in an online community of mothers. *Journal of Marketing Management*, 15 December. https://doi.org/10.1080/0267257x.2016.1156725

Carrington, B., & Bonnett, A. (1997). The other Canadian "mosaic": "Race" equity education in Ontario and British Columbia." *Comparative Education, 33*(3), 411–431. https://doi.org/10.1080/03050069728442

Castells, M. (1996). *The rise of the network society: The information age*, Vol. 1. Blackwell.

Castells, M. (1997 [2004]). *The power of identity:* The information age, Vol. 2 (2nd ed., 1997). Blackwell.

Castells, M. (1998). *End of millennium: The information age*, Vol. 3 (3rd ed., 2010). Blackwell.

Cayla, J., & Eckhart, G. M. (2008). Asian brands and the shaping of a transnational imagined vommunity. *Journal of Consumer Research, 35*(2), 216–230. https://doi.org/10.1086/587629

CBC. (2009). Estimated 2.8 million Canadians live abroad. *CBC.ca*, 29 October 2009.

Challinor, A. E. (2011). *Canada's immigration policy: A focus on human capital.* Migration Policy Institute.

Chapman, M. (1997). Social anthropology, business studies, and cultural issues. *International Studies of Management and Organization, 26*(4), 3–29. https://doi.org/10.1080/00208825.1996.11656692

Chapman, M., Clegg, J., & Gajewska-De Mattos, H. (2004). Poles and Germans: An international business relationship. *Human Relations, 57*(8), 983–1015. https://doi.org/10.1177/0018726704045837

Chen, R.-L. (2012). Beyond national identity in Taiwan: A multidimensional and evolutionary conceptualization. *Asian Survey, 52*(5), 845–871. https://doi.org/10.1525/as.2012.52.5.845

Chen, T.-J. (2004). Liability of foreignness and entry mode choice: Taiwanese firms in Europe. *Journal of Business Research, 59*, 288–294. https://doi.org/10.1016/j.jbusres.2005.04.009

Cheung, G. C. K. (2009). Made in China vs made by Chinese: Global identities of Chinese business." *Journal of Contemporary China, 18*(58), 1–5. https://doi.org/10.1080/10670560802431388

Chiang, L.-H. N. (2011). Return migration: The case of the 1.5 generation of Taiwanese in Canada and New Zealand. *China Review, 11*(2), 91–123.

Chiu, Y.-P., Wu, M., Zhuang, W.-L., & Hsu, Y.-Y. (2009). Influences on expatriate social networks in China. *The International Journal of Human Resource Management, 20*(4), 790–809. https://doi.org/10.1080/09585190902770703

Chow, P. C. Y. (2012). Introduction. In P. C. Y. Chow (Ed.), *National identities and economic interest: Taiwan's competing options and their implications for regional stability* (pp. 3–14). Palgrave.

Coe, N. M., Dicken, P., & Hess, M. (2008). Global production networks: Realizing the potential. *Journal of Economic Geography, 8*(3), 271–295. https://doi.org/10.1093/jeg/lbn002

Cohen, A. P. (1985). *The symbolic construction of community.* Tavistock.

Cohen, A. P. (1986). Of symbols and boundaries, or, does Ertie's greatcoat hold the key? In A .P. Cohen (Ed.), *Symbolising boundaries: Identity and diversity in British cultures* (pp. 1–22). Manchester University Press.

Cohen, A. P. (1994). *Self consciousness: An alternative ethnography of identity.* Routledge.

Cohen, R. (1997). *Global diasporas: An introduction.* University College London.

Cook-Gumperz, J. (1983). Socialization, social identity, and discourse." In A. Jacobson-Widding (Ed.), *Identity: Personal and socio-cultural, a symposium* (pp. 123–33). Almqvist and Wiksell International.

Cordero-Guzmán, H. R. (2005). Community-based organisations and migration in New York City. *Journal of Ethnic and Migration Studies, 31*(5), 889–909. https://doi.org/10.1080/13691830500177743

Courtney, C., & Thompson, P. (1996). *City lives: The changing voices of British finance.* Methuen.

Crowley-Henry, M., O'Connor, E., & Al Ariss, A. 2018. Portrayal of skilled migrants' careers in business and management studies: A review of the literature and future research agenda. *European Management Review, 15*, 375–394. https://doi.org/10.1111/emre.12072

Cruz-Malave, A. (Ed.). (2002). *Queer globalizations: Citizenship and the afterlife of colonialism.* NYU Press.

Dahles, H. (2005). Culture, capitalism, and political entrepreneurship: Transnational ventures of the Singapore-Chinese in China. *Culture and Organization, 11*(1), 45–58. https://doi.org/10.1080/14759550500062342

Danielsen, M. (2012). On the road to a common Taiwan identity. In P. C. Y. Chow (Ed.), *National identities and economic interest: Taiwan's competing options and their implications for regional stability* (pp. 135–52). Palgrave.

Davidson, J. W. (2007). *The island of Formosa, past and present: History, people, resources and commercial prospects* (Original work published 1903). Kessinger.

Dawley, E. N. (2009). The question of identity in recent scholarship on the history of Taiwan. *China Quarterly, 198*, 442–452. https://doi.org/10.1017/s030574100900040x

Dekker, R., & Engberson, G. (2014). How social media transform migrant networks and facilitate migration. *Global Networks, 14*(4), 401–418. https://doi.org/10.1111/glob.12040

Deng, F. J., Huang, L., Carraher, S., & Duan, J. (2009). International expansion of family firms: An integrative framework using Taiwanese manufacturers. *Academy of Entrepreneurship Journal 15*(1), 25–44.

Derudder, B., & Taylor, P. (2005). The cliquishness of world cities. *Global Networks, 5*, 71–91. https://doi.org/10.1111/j.1471-0374.2005.00108.x

Devinney, T. M., & Hartwell, C. A. (2020). Varieties of populism. *Global Strategy Journal, 10*, 32–66. https://doi.org/10.1002/gsj.1373

Dicken, P. (2010). *Global shift* (6th Ed.) Sage.

Dürrschmidt, J. (2006). So near yet so far: Blocked networks, global links, and multiple exclusion in the German–Polish borderlands. *Global Networks, 6*, 245–263. https://doi.org/10.1111/j.1471-0374.2006.00143.x

Dyck, D. (2016). Canada's Conservatives should watch the GOP – to see what they risk becoming. *Globe and Mail*, 27 May 2016. www.theglobeandmail.com/opinion/editorials/canadas-conservatives-should-watch-the-gop-to-see-what-they-must-never-become/article30190815/?click=sf_globefb

Ebner, D. (2009). A dynasty founded on instinct. *Globe and Mail*, 3 September 2009.

The Economist (US). (1995). Inheriting the bamboo network: The overseas Chinese. 23 December.

The Economist (US) (2015). No country for old men. 8 January.

The Economist (China). (2016a). Taiwan's Kuomintang party is broke and adrift. 15 December.

The Economist (China). (2016b). Taiwan fears becoming Donald Trump's bargaining chip. 27 December.

Elger, T., & Smith, C. (1994). *Global Japanization?: The transnational transformation of the labour process.* Routledge.

Eriksen, T. H. (1993). *Ethnicity and nationalism: Anthropological perspectives.* Pluto Press.

Falzon, M. A. (Ed.). (2016). *Multi-sited ethnography: Theory, praxis, and locality in contemporary research.* Routledge.

Fechter, A. (2007). *Transnational lives: Expatriates in Indonesia.* Ashgate.

Foner, N. (1997). *What's new about transnationalism? New York immigrants today and at the turn of the century*, paper presented at the Conference on Transnational Communities and the Political Economy of New York in the 1990s, The New School for Economic Research, 21–22 February 1997.

Forster, N. (2000). The myth of the "international manager." *International Journal of Human Resource Management, 11*(1), 126–142. https://doi.org/10.1080/095851900340024

Forsythe, D. (1989). German identity and the problem of history. In E. Tonkin et al. (Eds.), *History and Ethnicity*, Vol. 27 (pp. 137–156). Routledge.

Fox, R. G. (1977). *Urban anthropology. Cities in their cultural settings.* Prentice Hall.

Funnell, L. (2014). *Warrior women: Gender, race, and the transnational Chinese action star.* SUNY Press.

Gammeltoft-Hansen, T., and Sorensen, N. (Eds.). (2013). *The migration industry and the commercialization of international migration.* Routledge.

Gillespie, K., & McBride, J. B. (2013). Counterfeit smuggling: Rethinking paradigms of diaspora investment and trade facilitation. *Journal of International Management, 19*, 66–81. https://doi.org/10.1016/j.intman.2012.08.001

Glick Schiller, N., & Salazar, N. (2012). Regimes of mobility across the globe. *Journal of Ethnic and Migration Studies, 39*(2), 183–200. https://doi.org/10.1080/1369183x.2013.723253

Goffman, E. (1959). *The presentation of self in everyday life.* University of Edinburgh.

Goffman, E. (1961). *Encounters: Two studies in the sociology of interaction.* Bobbs-Merrill.

Goffman, E. (1963). *Behaviour in public places: Notes on the social organization of gatherings.* The Free Press of Glencoe.

Goffman, E. (1967). *Interaction ritual: Essays on face-to-face behaviour.* Doubleday.

Goh, R. B. H. (2014). The Lord of the Rings and New Zealand: Fantasy pilgrimages, imaginative transnationalism, and the semiotics of the (Ir) Real. *Social Semiotics, 24*(3), 263–282. https://doi.org/10.1080/10350330.2013.866781

Gonzales, C. (2019). Is the locus of class development of the transnational capitalist class situated within nation-states or in the emergent transnational space?" *Global Networks, 19*(2), 261–279. https://doi.org/10.1111/glob.12220

Greverus, I. M. (1979). *Auf der Suche nach Heimat.* Beck.

Guarnizo, L. E., & Smith, M. P. (1998). The locations of transnationalism. In M. P. Smith & L. E. Guarnizo (Eds.), *Transnationalism from below.*Comparative Urban and Community Research, Vol. 6 (pp. 13–34). Transaction Publishers, 13–34

Hall, B. (2003). Modernization and the social construction of national identity: The case of Taiwanese identity. *Berkeley Journal of Sociology, 47*, 135–169.

Hamori, M., & Koyuncu, B. (2011). Career advancement in large organizations in Europe and the United States: Do international assignments add value?" *International Journal of Human Resource Management, 22*(4), 843–862. https://doi.org/10.1080/09585192.2011.555128

Hannerz, U. (1992). *Cultural complexity: Studies in the social organization of meaning.* Columbia University Press.

Hannerz, U. (1996). *Transnational connections: Culture, people, places.* Routledge.

Harrison, M. (2007). *Legitimacy, meaning, and knowledge in the making of Taiwanese identity.* Palgrave.

Harvey, W. S. (2008). Strong or weak ties? British and Indian expatriate scientists finding jobs in Boston. *Global Networks, 8*(2), 453–473. https://doi.org/10.1111/j.1471-0374.2008.00234.x

He, H. H. (2012). "Chinesenesses" outside Mainland China: Macao and Taiwan through post 1997 Hong Kong cinema, *Culture Unbound, 4,* 297–325. https://doi.org/10.3384/cu.2000.1525.124297

Head, D. (1992). *"Made in Germany": The corporate identity of a nation.* Hodder and Stoughton.

Helbling, M., & Teney, C. (2015). The cosmopolitan elite in Germany: Transnationalism and postmaterialism. *Global Networks, 15,* 446–468. https://doi.org/10.1111/glob.12073

Held, D., McGrew, A., Goldblatt, D., & Perraton, J. (1999). *Global transformations: Politics, economics, and culture.* Polity Press.

Herbert, J. (2012). The British Ugandan Asian diaspora: Multiple and contested belongings. *Global Networks, 12*(3), 296–313. https://doi.org/10.1111/j.1471-0374.2012.00353.x

Hill, D. (2015). Immigration matters less in London but Ukip factor still counts. *The* Guardian, 3 April 2015.

Hirst, P. Q., & Thompson, G. (1996). *Globalization in question: The international economy and the possibilities of governance.* Polity Press.

Ho, E. L. (2009). Constituting citizenship through the emotions: Singaporean transmigrants in London. *Annals of the Association of American Geographers, 99*(4), 788–804. https://doi.org/10.1080/00045600903102857

Ho, E. L. (2011a). "Claiming" the diaspora: Elite mobility, sending state strategies, and the spatialities of citizenship. *Progress in Human Geography, 35*(6), 757–772. https://doi.org/10.1177/0309132511401463

Ho, E. L. (2011b). Migration trajectories of "highly skilled" middling transnationals: Singaporean transmigrants in London." *Population, Space, and Place, 17*(1), 116–129. https://doi.org/10.1002/psp.569

Hoang, L. A. (2016). Governmentality in Asian migration regimes: The case of labour migration from Vietnam to Taiwan. *Population, Space, and Place, 23*(3), 1–12. https://doi.org/10.1002/psp.2019

Hofstede, G. (1980). *Culture's consequences: International differences in work-related values.* SAGE.

Holmes, C. (1988) *John Bull's island: Immigration and British society, 1871–1971.*: Macmillan.

Holý, L. (1987). Introduction: Description, generalization, and comparison: Two paradigms. In L. Holý (Ed.), *Comparative anthropology* (pp. 1–21). Basil Blackwell.

Holý, L. (Ed.). (1987). *Comparative anthropology*. Basil Blackwell.

Horst, H. A. (2006). The blessings and burdens of communication: Cell phones in Jamaican transnational social fields." *Global Networks, 6*, 143–159. https://doi.org/10.1111/j.1471-0374.2006.00138.x

Ibbitson, J. (2014). Conservatives changed the nature of Canadian immigration. *Globe And Mail*, 15 December 2014.

Iyer, P. (2000). *The global soul: Jet lag, shopping malls, and the search for home*. Bloomsbury.

Jenkins, A. (1988). *The City: London's square mile*. Viking Kestrel.

Jenkins, R. (1996). *Social Identity*. Routledge.

Jiang, Q., & Fung, A. Y. H. (2019). Games with a continuum: Globalization, regionalization, and the nation-state in the development of China's online game industry. *Games and Culture, 14*(7–8), 801–824. https://doi.org/10.1177/1555412017737636

Kelley, N., & Trebilcock, M. (1998). *The making of the mosaic: A history of Canadian immigration policy*. University of Toronto Press.

Knight, G. A., & Cavusgil, S. T. (2004). Innovation, organizational capabilities, and the born-global firm. *Journal of International Business Studies, 35*, 334–334. https://doi.org/10.1057/palgrave.jibs.8400096

Koch, I. (2017). What's in a vote?: Brexit beyond culture wars. *American Ethnologist 44*(2), 225–230. https://doi.org/10.1111/amet.12472

Koopmans, R., & P. Statham. (1999). Challenging the liberal nation-state? Postnationalism, multiculturalism, and the collective claims making of migrants and ethnic minorities in Britain and Germany." *The American Journal of Sociology, 105*(3), 652–696. https://doi.org/10.1086/210357

Kozinets, R. V. (2009). *Netnography: Doing ethnographic research online*. SAGE.

Krätke, S. (2014). How manufacturing industries connect cities across the world: Extending research on "multiple globalizations." *Global Networks, 14*, 121–147. https://doi.org/10.1111/glob.12036

Kshetri, N., & Alcantara, L. L. (2015). Cyber-threats and cybersecurity challenges: A cross-cultural perspective." In N. Holden, S. Michaelova, & S. Tietze (Eds.), *The Routledge companion to cross-cultural management* (pp. 285–293). Routledge.

Lan, P.-C., & Wu, Y.-F. (2016). Exceptional membership and liminal space of identity: Student migration from Taiwan to China. *International Sociology, 31*(6), 742–763. https://doi.org/10.1177/0268580916662389

Landolt, P., & Goldring, L. (2010). Political cultures and transnational social fields: Chileans, Colombians, and Canadian activists in Toronto. *Global Networks, 10*, 443–466. https://doi.org/10.1111/j.1471-0374.2010.00290.x

Lewis, M. (1989). *Liar's poker: Two cities, true greed*. Hodder and Stoughton.

Leyshon, A., & Thrift, N. (1997). *Money/space: Geographies of monetary transformation*. Routledge.

Li, C. (1994). University networks and the rise of Qinghua graduates in China's leadership. *The Australian Journal of Chinese Affairs, 32*, 1–30. https://doi.org/10.2307/2949825

Lien, P.-T. (2011). Chinese American attitudes toward homeland government and politics: A comparison among immigrants from China, Taiwan, and Hong Kong. *Journal of Asian American Studies, 14*(1), 1–31. https://doi.org/10.1353/jaas.2011.0007

Lin, P. (2011). Chinese diaspora "at home": Mainlander Taiwanese in Dongguan and Shanghai. *China Review, 11*(2): 43–64.

Liu, M., & Hung, L-C. (2002). Identity issues in Taiwan's history curriculum. *International Journal of Educational Research 37*, 567–586. https://doi.org/10.1016/s0883-0355(03)00051-x

Lo, S-H. (2002). Diaspora regime into nation: Mediating hybrid nationhood in Taiwan. *Javnost – The Public, 9*(1): 65–83. https://doi.org/10.1080/13183222.2002.11008794

Lynch, T. G. (2009). "A kindred and congenial element": Irish-American nationalism's embrace of Republican rhetoric. *New Hibernia Review, 13*(2), 77–91. https://doi.org/10.1353/nhr.0.0073

MacAskill, A., & Davies, A. (2017). Banks begin London exodus as hopes of transitional deal fade. *Reuters*, 11 July 2017.

Marlowe, J. (2019). Refugee resettlement, social media, and the social organization of difference. *Global Networks*, https://doi.org/10.1111/glob.12233

Massey, D. (1991). A global sense of place. *Marxism Today, 38*, 24–29.

Mauss, M. (1934). Les techniques du corps. *Journal de Psychologie, 32*(3–4), 271–293. https://doi.org/10.1522/cla.mam.tec

McDowell, L. M. (1997). A Tale of two cities? Embedded organizations and embodied workers in the City of London. In R. Lee & J. Willis (Eds.), *Geographies of Economies* (pp. 118–129). Arnold.

McSweeney, B. (2002). Hofstede's model of national cultural differences and their consequences: A triumph of faith – a failure of analysis. *Human Relations, 55*(1), 89–118. https://doi.org/10.1177/0018726702551004

McSweeney, B. (2013). Fashion founded on a flaw: The ecological mono-determinist fallacy of Hofstede, GLOBE, and followers. *International Marketing Review, 30*(5), 483–504. https://doi.org/10.1108/imr-04-2013-0082

Meinhof, U. (2001). *Changing nations, states, and identities in German–Polish border regions.* Paper presented at ESRC Transnational Communities Seminar, Oxford, 25 January 2001.

Miller, D., & Slater, D. (2000). *The Internet: An ethnographic approach.* Berg.

Mirchandani, K. (2004). Practices of global capital: Gaps, cracks, and ironies in transnational call centres in India. *Global Networks, 4*, 355–373. https://doi.org/10.1111/j.1471-0374.2004.00098.x

Montgomery, A. F. (2008). Virtual enclaves: The influence of alumni email lists on the workspaces of transnational software engineers. *Global Networks, 8*, 71–93. https://doi.org/10.1111/j.1471-0374.2008.00186.x

Moore, F. (2005). *Transnational business cultures: Life and work in a multinational corporation*. Ashgate.

Moore, F. (2006). Strategy, power, and negotiation: Social control and expatriate managers in a German multinational corporation. *International Journal of Human Resource Management, 17*(3), 399–413. https://doi.org/10.1080/09585190500521359

Moore, F. (2016). City of sojourners versus city of settlers: Transnationalism, location, and identity among Taiwanese professionals in London and Toronto. *Global Networks 16*(3), 372–390. https://doi.org/10.1111/glob.12120

Moya, J. C. (2005). Immigrants and associations: A global and historical perspective. *Journal of Ethnic and Migration Studies, 31*(5), 833–864. https://doi.org/10.1080/13691830500178147

Murray, S. O., & Hong, K. (1991). American anthropologists looking through Taiwanese culture. *Dialectical Anthropology, 13*(3/4), 273–299. https://doi.org/10.1007/bf00301241

Muyard, F. (2012). Taiwanese national identity, cross-strait economic interaction, and the ingegration paradigm. In P. C. Y. Chow (Ed.), *National identities and economic interest: Taiwan's competing options and their implications for regional stability* (pp. 153–186). Palgrave.

Ndhlovu, F. (2016). A decolonial critique of diaspora identity theories and the notion of superdiversity. *Diaspora Studies, 9*(1), 28–40. https://doi.org/10.1080/09739572.2015.1088612

Neal, Z. P. (2008). The duality of world cities and firms: Comparing networks, hierarchies, and inequalities in the global economy. *Global Networks, 8*, 94–115. https//doi.org/10.1111/j.1471-0374.2008.00187.x

Nonini, D., & Ong, A. (1997). Chinese transnationalism as an alternative modernity. In A. Ong & D. Nonini (Eds.), *Ungrounded empires: The cultural politics of modern Chinese transnationalism* (pp. 1–33). Routledge.

Numazaki, I. (1986). Networks of Taiwanese big business: A preliminary analysis. *Modern China, 12*(4), 487–534. https://doi.org/10.1177/009770048601200403

Ong, A. (1999). *Flexible citizenship: The cultural logics of transnationality*. Duke University Press.

Park, H. H., & Luo, Y. (2001). *Guanxi* and organizational dynamics: Organizational networking in Chinese firms. *Strategic Management Journal, 22*(5), 455–77. https://doi.org/10.1002/smj.167

Parnreiter, C. (2013). The global city tradition. In M. Acuto & W. Steele (Eds.), *Global city challenges: Debating a concept, improving the practice* (pp. 15–32). Palgrave Macmillan.

Portes, A. (1998). *Globalisation from below: The rise of transnational communities*, ESRC Transnational Communities Programme Working Paper.

Preibisch, J., & Hennebry, K. (2011). Temporary migration, chronic effects: The health of international migrant workers in Canada. *Canadian Medical Association Journal, 183*(9), 1033–1038. https://doi.org/10.1503/cmaj.090736 .Medline:21502343

Redclift, V. M., & Rajina, F. B. (2019). The hostile environment, Brexit, and "reactive" or "protective transnationalism." *Global Networks*. https://doi.org /10.1111/glob.12275f.

Rehnberg, M., & Ponte, S. (2018). From smiling to smirking? 3D printing, upgrading, and the restructuring of global value chains. *Global Networks 18*(1), 57–80. https://doi.org/10.1111/glob.12166

Robertson, R. (1992). *Globalization: Social theory and global culture*. SAGE.

Rogers, A. (2005). Observations on transnational urbanism: Broadening and narrowing the field. *Journal of Ethnic and Migration Studies, 31*(2), 403–407. https://doi.org/10.1080/1369183042000339990

Rubin, J. P. (2016). London risks losing its status as the capital of the world. *The Economist*, 4 March 2016. www.economist.com/britain/2016/03/04/london -risks-losing-its-status-as-the-capital-of-the-world

Ryan, L., & Mulholland, J. (2014). French connections: The networking strategies of French highly skilled migrants in London. *Global Networks, 14*(2), 148–166. https://doi.org/10.1111/glob.12038

Salaff, J., Grieve, A., & Ping, L. X. L. (2002). Paths into the economy: Structural barriers and the job hunt for skilled PRC migrants in Canada. *International Journal of Human Resource Management, 13*(3), 450–464. https://doi.org/10 .1080/09585190110111477

Sanday, P. R. (1979). The ethnographic paradigm(s). *Administrative Science Quarterly, 24*(4), 527–538. https://doi.org/10.2307/2392359

Sassen, S. (2001). *The global city: New York, London, Tokyo* (2nd ed.). Princeton University Press.

Schrover, M., & Vermeulen, F. (2005). Immigrant organisations. *Journal of Ethnic and Migration Studies, 31*(5), 823–832. https://doi.org/10.1080 /13691830500177792

Schubert, G. (2004). Taiwan's political parties and national identity: The rise of an overarching consensus. *Asian Survey, 44*(4), 534–554. https://doi.org /10.1525/as.2004.44.4.534

Selmer, J. (2002). The Chinese connection?: Adjustment of Western vs. overseas Chinese expatriate managers in China. *Journal of Business Research, 55*, 41–50. https://doi.org/10.1016/s0148-2963(00)00132-6

Sharma, S. (1982). East Indians and the Canadian ethnic mosaic: An overview. *South Asia Bulletin, 2*(1): 6–18. https://doi.org/10.1215/07323867-2-1-6

Shaw, A. (1988). *A Pakistani community in Britain*. Blackwell.

Shimoda, Y. (2017). *Transnational organizations and cross-cultural workplaces.* Palgrave.

Sklair, L. (2001). *The transnational capitalist class.* Blackwell.

Smith, A. D. (1995). *Nations and nationalism in a global era.* Polity Press

Smith, M. P. (2001). *Transnational urbanism: Locating globalization.* Blackwell.

Smith, M. P. (2010). Transnational urbanism revisited. *Journal of Ethnic and Migration Studies, 31*(2), 235–244. https://doi.org/10.1080/1369183042000339909

Somerville, W., Sriskandarajah, D., & Latorre, M. (2009). United Kingdom: A reluctant country of immigration. *Migration Information Source: The Online Journal of the Migration Policy Institute.* www.migrationpolicy.org.

Spence, L. J. (1998). *Comparative European business ethics: A comparison of the ethics of the recruitment interview in Germany, the Netherlands, and the UK, using Erving Goffman's frame analysis.* PhD diss., Brunel University.

Sperber, D. (1974). *Rethinking symbolism.* Cambridge University Press.

Stratham, P. (2003). Understanding anti-asylum rhetoric: Restrictive politics or racist publics? *The Political Quarterly, 74,* 163–177. https://doi.org/10.1111/j.1467-923x.2003.00588.x0

Strecker, I. (1988). *The social practice of symbolization: An anthropological analysis.* Athlone Press.

Sun, K. C.-Y. (2014). Transnational healthcare seeking: How ageing Taiwanese return migrants view homeland public benefits. *Global Networks, 14,* 533–550. https://doi.org/10.1111/glob.12050

Tajdin, B. (2013). Will Iran's national internet mean no world wide web? *BBC News Online,* 27 April 2013. https://www.bbc.co.uk/news/world-middle-east-22281336

Tanzer, A. (1994). The bamboo network. *Forbes, 154*(2), 138–145.

Teo, S. Y. (2011). "The moon back home is brighter"?: Return migration and the cultural politics of belonging. *Journal of Ethnic and Migration Studies, 37*(5), 805–820. https://doi.org/10.1080/1369183x.2011.559720

Thompson, E. C. (2009). Mobile phones, communities, and social networks among foreign workers in Singapore. *Global Networks, 9,* 359–380. https://doi.org/10.1111/j.1471-0374.2009.00258.x

Thrift, N. (1994). On the social and cultural determinants of international financial centres: The case of the City of London. In S. Corbridge, R. Martin, & N. Thrift (Eds.), *Money, power, and space* (pp. 327–355). Blackwell.

Tien, C., & Luan, C.-J. (2015) Is the magic of the diaspora fact or fiction? A study of Taiwan' s trade performance in the bamboo network. *Emerging Markets Finance and Trade, 51,* 234–250. https://doi.org/10.1080/1540496x.2014.998890

Tomlinson, J. (1999). *Globalization and culture.* Polity Press.

Tsang, A. K. T., Irving, H., Alaggia, R., Chau, S. B. Y., & Benjamin, M. (2003). Negotiating ethnic identity in Canada: The case of the "satellite children.'" *Youth and Society, 34*(3), 359–384. https://doi.org/10.1177/0044118x02250124

Tseng, Y.-C. (2017). Should I stay or should I go? Migration trajectories of Chinese-Taiwanese couples in third countries. *Asia and Pacific Migration Journal, 26*(4), 413–435. https://doi.org/10.1177/0117196817747296

Tsui, A. (2016). Reflections on the so-called value-free ideal: A call for responsible science in the business schools. *Cross-Cultural Management and Strategy, 23*(1), 4–28. https://doi.org/10.1108/ccsm-08-2015-0101

Van Veen, K. (2018). How did the financial crisis affect the transnationality of the global financial elite? One step forward and one step back. *Global Networks, 18*(1), 105–126. https://doi.org/10.1111/glob.12182

Vertovec, S. (1999). Conceiving and researching transnationalism. *Ethnic and Racial Studies, 22*(2): 447–462. https://doi.org/10.1080/014198799329558

Vertovec, S. (2001). *Transnational social formations: Towards conceptual cross-fertilization*, paper presented at Workshop on Transnational Migration: Comparative Perspectives, Princeton University, 30 June 30–1 July 2001.

Vomiero, J., & Russell, A. (2019). Ipsos poll shows Canadians have concerns about immigration. Here are the facts. *Global News*, 4 January 2019. https://globalnews.ca/news/4794797/canada-negative-immigration-economy-ipsos

Wang, C.-H. (2003). Taipei as a global city: A theoretical and empirical examination." *Urban Studies, 40*, 309–334. https://doi.org/10.1080/0042098022008029l

Wang, H.-L. (2004). National culture and its discontents: The politics of heritage and language in Taiwan, 1949–2003. *Comparative Studies in Society and History, 46*(4), 786–815. https://doi.org/10.1017/s0010417504000362

Wang, L. (2007). Diaspora, identity, and cultural citizenship: The Hakkas in "multicultural Taiwan." *Ethnic and Racial Studies, 30*(5), 875–895. https://doi.org/10.1080/01419870701491861

Warrell, H. (2015). Ease foreign graduate visa restrictions, business leaders urge. *Financial Times*, 22 February 2015.

Waters, J. L. (2006). Geographies of cultural capital: Education, international migration and family strategies between Hong Kong and Canada. *Transactions of the Institute of British Geographers*, New Series, *31*(2), 179–192. https://doi.org/10.1111/j.1475-5661.2006.00202.x

Waters, M. (1995). *Globalization*. Routledge.

Watson, A. (1995). *The Germans: Who are they now?* Methuen.

Weale, S. (2015). London's international students prove lucrative for UK economy, claims study. *The Guardian*, 18 May 2015.

Weidenfeld, W., & Korte, K.-R. (1991). *Die Deutschen: Profil einer Nation*. Klett-Cotta.

Weigelt, K. (Ed.). (1984). *Heimat und Nation: zur Geschichte und Identität der Deutschen*. Von Hase und Köhler.

White, J. (1997). Turks in the new Germany. *American Anthropologist, 99*(4), 754–769. https://doi.org/10.1525/aa.1997.99.4.754

Wilentz, R. S. (1979). Industrializing America and the Irish: Towards the new departure. *Labor History, 20*(4), 579–595. https://doi.org/10.1080/00236567908584555

Wong, A. L. Y., Shaw, G. H., & Ng., D. K. C. (2010). Taiwan Chinese managers' personality: Is Confucian influence on the wane?" *International Journal of Human Resource Management, 21*(7), 1108–1123. https://doi.org/10.1080/09585191003783546

Wu, T.-M. (2004). "Economic history of Taiwan: A survey." *Australian Economic History Review, 44*(3), 294–306. https://doi.org/10.1111/j.1467-8446.2004.00123.x

Xin, K. R., & Pearce, J. L. (1996). *Guanxi*: Connections as substitutes for formal institutional support. *The Academy of Management Journal, 39*(6), 1641–1658. https://doi.org/10.2307/257072

Yang, D. M.-H., & Chang, M.-K. (2010). Understanding the nuances of *waishengren*: History and agency. *China Perspectives, 3*(83), 108–122. https://doi.org/10.4000/chinaperspectives.5310

Yang, M. (1994). *Gifts, favors and banquets: The art of social relationships in China.* Cornell University Press.

Yeoh, B. S. A. (2005). Observations on transnational urbanism: Possibilities, politics, and costs of simultaneity. *Journal of Ethnic and Migration Studies, 31*(2), 409–413. https://doi.org/10.1080/1369183042000339006

Yeoh, B. S. A., & Huang, S. (2011). Introduction: Fluidity and friction in talent migration. *Journal of Ethnic and Migration Studies, 37*(5), 681–690. https://doi.org/10.1080/1369183x.2011.559710

Yeoh, B. S. A., & Willis, K. (2011). Singaporean and British transmigrants in China and the cultural politics of "contact zones." *Journal of Ethnic and Migration Studies, 31*(2), 269–285. https://doi.org/10.1080/1369183042000339927

Yu, J., & Bairner, A. (2008). Proud to be Chinese: Little League Baseball and national identities in Taiwan during the 1970s." *Identities, 15*(2), 216–239. https://doi.org/10.1080/10702890801904636

Yu, Y.-S. (2011). "Zhong Hua": The origin and change of the ideas about "Zhong Hua." *Journal of the History of Ideas in East Asia, 11*, 327–356.

Zhong, Z.-J. (2011). The effects of collective MMORPG (massively multiplayer online role-playing games) play on gamers' online and offline social capital. *Computers in Human Behavior, 27*(6), 2352–2363. https://doi.org/10.1016/j.chb.2011.07.014

Zhou, Y., & Hsu, J.-Y. (2011). Divergent engagements: Roles and strategies of Taiwanese and mainland Chinese returnee entrepreneurs in the IT industry." *Global Networks, 11*, 398–419. https://doi.org/10.1111/j.1471-0374.2010.00302.x

Index

Lightning Source UK Ltd.
Milton Keynes UK
UKHW011306090223
416759UK00019B/311/J